Igor Gilitschenski

Deterministic Sampling for Nonlinear Dynamic State Estimation

Karlsruhe Series on Intelligent Sensor-Actuator-Systems

Volume 18

ISAS | Karlsruhe Institute of Technology
Intelligent Sensor-Actuator-Systems Laboratory

Edited by Prof. Dr.-Ing. Uwe D. Hanebeck

Deterministic Sampling for Nonlinear Dynamic State Estimation

by
Igor Gilitschenski

Scientific Publishing

Dissertation, Karlsruher Institut für Technologie (KIT)
Fakultät für Informatik, 2015

Impressum

Karlsruher Institut für Technologie (KIT)
KIT Scientific Publishing
Straße am Forum 2
D-76131 Karlsruhe

KIT Scientific Publishing is a registered trademark of Karlsruhe Institute of Technology. Reprint using the book cover is not allowed.

www.ksp.kit.edu

Print on Demand 2016

ISSN 1867-3813
ISBN 978-3-7315-0473-3
DOI 10.5445/KSP/1000051670

Deterministic Sampling for Nonlinear Dynamic State Estimation

zur Erlangung des akademischen Grades eines

Doktors der Ingenieurwissenschaften

von der Fakultät für Informatik
des Karlsruher Instituts für Technologie (KIT)

genehmigte

Dissertation

von

Igor Gilitschenski

aus Orhei

Tag der mündlichen Prüfung: 24.4.2015

Erster Gutachter: Prof. Dr.-Ing. Uwe D. Hanebeck

Zweiter Gutachter: Simon J. Julier, PhD

Acknowledgment

This work presents the results of my PhD research, which was carried out at the Intelligent Sensor-Actuator-Systems (ISAS) laboratory within the Institute for Anthropomatics and Robotics (IAR) of the Karlsruhe Institute of Technology (KIT). For over three years I benefited from the support of many people all of whom contributed to the success of my PhD.

First, I am grateful for having had Uwe D. Hanebeck as my main supervisor. He gave me the opportunity to work in his lab and motivated me by posing interesting and challenging research questions. My stay at the University College London (UCL) was an unforgettable experience thanks to the collaboration with my co-supervisor Simon J. Julier. His endless curiosity and constant flow of brilliant ideas were always inspiring.

My work was supported by the German Research Foundation (DFG) as part of the Research Training Group 1194 (GRK 1194) on "Self-organizing Sensor-Actuator-Networks" and a travel grant for my visit to London by the Karlsruhe House of Young Scientists (KHYS). Apart from the financial support, the meetings and discussions with colleagues from the Research Training Group were very insightful in gaining a broader view on our research.

I was blessed with great colleagues from diverse backgrounds all of whom made my work at ISAS and my time in Karlsruhe a memorable experience. The collaborations with Gerhard Kurz, Jannik Steinbring, and Maxim Dolgov were intellectually stimulating and very fruitful. Christof Chlebek has been a great office mate with whom I had a great time and many discussions about our research, academia, and life in general. Many thanks go to the students working with me on different projects for their curiosity and dedication.

There are numerous people outside of academia who deserve my gratitude. Particularly, I am deeply grateful to my parents, grandparents, and my sister Alina for providing a lifetime of wisdom and encouragement. I was very lucky to have great friends for whom I had way to little time during my PhD studies. Last, but by no means least, my girlfriend Svenja deserves most of my gratitude for her love and support. She always believed in me and my ambitions.

Zurich, March 2016 Igor Gilitschenski

Contents

Notation

General Conventions

x	Scalar
$\underline{x}$	Vector
$\underline{x}_{1:k}$	Set containing all vectors $\underline{x}_1$, ..., $\underline{x}_k$
$\underline{x}^\top$	Transposed vector
$\underline{0}$	Zero vector
$\mathbf{A}$	Matrix
$\mathbf{A}^\top$	Transposed matrix
$\mathbf{I}$	Identity matrix
$\text{tr}(\mathbf{A})$	Trace of matrix $\mathbf{A}$
$\det(\mathbf{A})$	Determinant of matrix $\mathbf{A}$
$\text{diag}(a_1, ..., a_n)$	Diagonal matrix with entries a_1, ... a_n
$\mathbb{N}$	Set of all natural numbers
$\mathbb{R}$	Set of all real numbers
S^n	Set of all unit vectors in $\mathbb{R}^{n+1}$
$\text{Bingham}(\mathbf{M}, \mathbf{Z})$	Bingham distribution with location / axes parameter $\mathbf{M}$ and concentration parameter $\mathbf{Z}$
$\mathcal{N}(\underline{\mu}, \mathbf{\Sigma})$	Gaussian distribution with mean $\underline{\mu}$ and covariance matrix $\mathbf{\Sigma}$
$\delta(\cdot)$	Dirac delta function
$\mathbb{E}(\underline{x})$	Expected value of random vector $\underline{x}$
$d_{w,K}(P_1, P_2)$	LCD distance with kernel $K(.)$ and weighting function $w(\cdot)$

$d_{KL}(P_1, P_2)$	Kullback-Leibler divergence of P_2 from P_1
$W_p(P_1, P_2)$	Wasserstein metric between P_1 and P_2
$\mathcal{L}_K[f]$	LCD transform of f using kernel K
$\mathcal{W}_a[f]$, $\mathcal{W}[f]$	(Generalized) Weierstrass transform of f
$\underline{a}^*$	Quaternion conjugation
$\underline{a}^\ddagger$	Conjugation of the quaternion base elements within the dual quaternion $\underline{a}$.
$\underline{a}^\dagger$	Conjugation of the dual unit ε within the dual quaternion $\underline{a}$
$\underline{a} \oplus \underline{b}$	Product of quaternions $\underline{a}$ and $\underline{b}$
$\underline{a} \boxplus \underline{b}$	Product of dual quaternions $\underline{a}$ and $\underline{b}$

Conventions in Filtering Algorithms

$a(\cdot)$, $\mathbf{A}$	System model / system matrix
$h(\cdot)$, $\mathbf{H}$	Measurement model / measurement matrix
$\underline{v}_t$	Measurement noise at time t
$\underline{w}_t$	System noise at time t
$\underline{x}_t$	True system state at time t
$\underline{z}_t$	Measurement at time t
f_t^p	Estimated density after prediction at time t
f_t^e	Estimated density after measurement update at time t
$\underline{x}_t^p$	Random variable distributed according to f_t^p
$\underline{x}_t^e$	Random variable distributed according to f_t^e
$\hat{\underline{x}}_t^p$	State estimate after prediction at time t
$\hat{\underline{x}}_t^e$	State estimate after measurement update at time t
$\hat{\underline{z}}_t$	Predicted measurement at time t
$\mathbf{C}_v$, $\mathbf{C}_w$	Covariance of measurement / system noise
$\hat{\mathbf{C}}_t^p$, $\hat{\mathbf{C}}_t^e$	Estimate of covariance after prediction / measurement update at time t

$\hat{\mathbf{M}}_t^p$, $\hat{\mathbf{M}}_t^e$	Estimate of Bingham location parameter matrix after prediction / measurement update at time t
$\hat{\mathbf{P}}_t^p$, $\hat{\mathbf{P}}_t^e$	Estimate of $SE(2)$ distribution parameter after prediction / measurement update at time t
$\hat{\mathbf{Z}}_t^p$, $\hat{\mathbf{Z}}_t^e$	Estimate of Bingham concentration parameter matrix after prediction / measurement update at time t

Abbreviations

CLT	Central limit theorem
CSB	Cauchy-Schwarz-Bunyakovsky inequality
EKF	Extended Kalman filter
IMU	Inertial measurement unit
i.i.d.	Independent and identically distributed
LCD	Localized cumulative distribution
LRKF	Linear regression Kalman filter
MSE	Mean square error
RBF	Radial basis function
RMSE	Root mean square error
SLAM	Simultaneous localization and mapping
UKF	Unscented Kalman filter

Zusammenfassung

Das Ziel der vorliegenden Arbeit ist die Verbesserung bestehender und Entwicklung neuer Algorithmen für die nichtlineare dynamische Zustandsschätzung. Die wichtigsten Beiträge sind neue Verfahren, um stetige Wahrscheinlichkeitsverteilungen durch diskrete Verteilungen zu approximieren. Dabei werden lineare und periodische zugrundeliegende Zustandsräume betrachtet. Für den linearen Fall schlagen wir ein neues Verfahren zur Approximation von gaußschen Dichten vor, welches ein globales Distanzmaß verwendet. Dieses kann stetige und diskrete Verteilungen simultan berücksichtigen. Es basiert auf lokalisierten kumulativen Verteilungen, einer Alternative zur Verteilungsfunktion, welche alle möglichen lokalen Wahrscheinlichkeitsmassen um einen vorgegebenen Punkt beschreibt. Die aus der Minimierung dieses Distanzmaßes resultierende diskrete Verteilung wird für approximative Integration verwendet und es wird gezeigt, dass die dadurch erhaltenen Ergebnisse aktuelle Verfahren übertreffen.

Im Falle periodischer Zustandsräume wird in dieser Arbeit ein deterministisches Sampling-Schema für die Bingham-Verteilung vorgeschlagen, die eine antipodal symmetrische Verteilung auf der n-dimensionalen Hyperkugel ist. Dieser Zustandsraum ist aufgrund der Möglichkeit Winkel und Orientierungen darzustellen von besonderem Interesse. Einheitsvektoren in 2D werden als Winkel aufgefasst, während Einheitsvektoren in 4D als Einheitsquaternionen und damit Orientierungen aufgefasst werden. Das Sampling-Verfahren beruht auf der Momentenmethode und kann als Hyperkugel-Äquivalent zum deterministischen Sampling im Unscented Kalman Filter (UKF) aufgefasst werden. Somit ist die Anwendung nicht nur auf die Bingham Verteilung beschränkt, sondern auch bei anderen antipodal symmetrischen Verteilungen auf der Hyperkugel möglich. Auch hier werden die deterministischen Sampling-Verfahren für approximative Integration verwendet. Dies erweitert und verbessert den Stand der Forschung auf zwei Arten. Zum einen ist es das erste Mal, dass ein deterministisches Sampling-Verfahren für die Hyperkugel vorgeschlagen wurde. Zum anderen wird nun die Betrachtung hoher Unsicherheiten möglich, indem die übliche Annahme lokaler Linearität des zugrundeliegenden Raumes vermieden wird.

Des Weiteren werden das deterministische Sampling auf linearen Zustandsräumen und Hyperkugeln für das Schätzen von Bewegungen starrer Körper in der Ebene kombiniert. Eine multiplikative Untergruppe dualer Quaternionen wird hergeleitet, um diese Art von Bewegungen darstellen zu können. Um die Unsicherheit von Bewegungen starrer Körper in der Ebene beschreiben zu können wird eine neue Verteilung vorgeschlagen. Durch diese Verteilung können Abhängigkeiten zwischen Position und Orientierung berücksichtigt werden. Diese Arbeit enthält die Herleitung von Parameterschätzern und ein effizientes Verfahren zur Berechnung der Normierungskonstante, welches auf einer Beziehung zur Bingham Verteilung beruht. Es wird gezeigt, wie deterministische Samples dieser neuen Verteilung aus deterministischen Samples der Bingham- und der Gauß-Verteilung berechnet werden können.

Die resultierenden Verfahren können unmittelbar auf nichtlineare dynamische Zustandsschätzung angewandt werden. Im Fall linearer Zustandsräume kann die vorgeschlagene Approximation der Gauß Verteilung in einem Linear Regression Kalman Filter (LRKF) eingesetzt werden und somit seine Qualität durch bessere numerische Propagation unsicherer Größen verbessern. Auf periodischen Bereichen führt das vorgeschlagene Sampling-Schema zu einem neuen Filter auf Grundlage der Bingham-Verteilung. Der vorgeschlagene Schätzalgorithmus hat im Falle schwachen Rauschens eine ähnliche Qualität wie aktuelle Verfahren welche üblicherweise auf der Annahme lokaler Linearität beruhen und übertrifft diese in beträchtlichem Maße, wenn starkes System- und/oder Messrauschen auftreten. Die neue Verteilung für ebene Bewegungen starrer Körper und das dazugehörige deterministische Sampling-Schema werden ebenfalls für die Herleitung eines neuen Filter genutzt, welches Position und Orientierung aus verrauschten Messungen schätzen kann. Dieses Filter hat ähnliche Vorteile, wie das vorgeschlagene Verfahren zur Orientierungsschätzung aufgrund ähnlicher Eigenschaften der zugrundeliegenden Verteilungen. Der Hauptvorteil dieser neu vorgeschlagenen Verfahren liegt darin begründet, dass sie gute Ergebnisse sowohl im Falle von hohem als auch im Falle von geringem Rauschen erzielen. Deswegen erleichtert der Beitrag dieser Arbeit den

Einsatz günstiger Sensoren und die simultane Verarbeitung qualitativ unterschiedlicher Daten in vielen realen Anwendungen, wie beispielsweise Wahrnehmung in der Robotik, intelligente Mobilität und der augmentierten sowie virtuellen Realität.

Abstract

The goal of this thesis is improving existing and suggesting novel filtering algorithms for nonlinear dynamic state estimation. Its main contributions are novel techniques for approximating continuous probability distributions by discrete distributions defined on the same continuous domain. It considers both, linear and periodic underlying state spaces. In the linear case, we propose a method for approximating Gaussian densities by using a global distance measure capable of handling continuous and discrete probability distributions simultaneously. It is based on the Localized Cumulative Distribution (LCD), an alternative to the cumulative distribution function that considers all possible local probability masses around a given point. The sample set obtained from minimizing the distance measure is applied to approximate integration, where the results are shown to outperform state-of-the-art approaches.

In case of periodic state spaces, this thesis proposes a deterministic sampling scheme for the Bingham distribution, which is an antipodally symmetric distribution defined on an N-dimensional hypersphere. This state space is of particular interest due to its capability of representing angles and orientations. Unit vectors in 2D are interpreted as angles, whereas unit vectors in 4D are interpreted as unit quaternions and, thus, are a suitable representation of uncertain orientations. The sampling scheme is based on moment matching and can be thought of as a hyperspherical equivalent to the samples obtained in the Unscented Kalman filter (UKF). Thus, its applicability is not restricted to the Bingham case but it is more broadly applicable to other antipodally symmetric hyperspherical distributions. Again, the resulting set of deterministically computed samples is applied to approximate integration. This contributes and improves the current state-of-the-art in two ways. First, it is the first time that a deterministic sampling scheme for the hypersphere has been proposed. Second, consideration of large uncertainties is now possible, because the typical exploitation of local linearity of the underlying domain is avoided.

Furthermore, deterministic sampling on linear domains and on hyperspheres is combined for estimation of planar rigid-body motions, which is an inference problem involving both directional and linear quantities. A multiplicative subgroup of dual quaternions for representing this type of motions is derived. A novel probability distribution is proposed in order to represent uncertain planar rigid-body motions. The proposed distribution is capable of describing dependencies between uncertain position and orientation. The contribution involves a derivation of parameter estimation techniques and an efficient method for computing the normalization constant that is based on a relationship with the Bingham distribution. It is shown how computing deterministic samples that approximate this distribution can be carried out using deterministic samples from the Bingham and the Gaussian distributions.

The resulting methods can be immediately applied to nonlinear dynamic state estimation. In case of linear state spaces, the proposed approximation of a Gaussian distribution can be used directly in a Linear Regression Kalman filter (LRKF) and, thus, improve its performance due to better numerical propagation of uncertain quantities. For periodic domains, the proposed deterministic sampling scheme gives rise to a new filter based on the Bingham distribution. The resulting estimation algorithm has similar performance to state of the art filtering techniques (which are usually based on assuming local linearity and then applying the UKF) in the case of weak noise, but significantly outperforms these techniques whenever strong system or measurement noise is present. The newly proposed distribution for planar rigid-body motions and its deterministic sampling scheme are also used for developing a new filter capable of estimating position and orientation from noisy measurements. It has similar advantages as the proposed orientation estimation procedure due to similarities between the underlying distributions. The main advantage of these newly proposed methods is the fact, that they are simultaneously capable of achieving good performance in both scenarios, those involving low noise and those involving high noise. Consequently, the contribution of this thesis facilitates the use of cheaper sensors and simultaneous handling of different sensor qualities in many real-world applications such as robotic perception, intelligent mobility, and mixed- and augmented reality.

CHAPTER 1

Introduction

"I believe in evidence. I believe in observation, measurement, and reasoning, confirmed by independent observers. I'll believe anything, no matter how wild and ridiculous, if there is evidence for it. The wilder and more ridiculous something is, however, the firmer and more solid the evidence will have to be."
- Isaac Asimov

The recent years have witnessed a massive deployment of different kinds of sensors in consumer goods. On the one hand, this development was driven by advances in sensor production, e.g., Microelectromechanical Systems (MEMS) pushing the limits towards ever cheaper, more precise, and less energy consuming sensors. On the other hand, the demand for better perception within products such as smartphones, cars, and novel robotic applications has created massive market opportunities for these sensors. Practical applications, e.g., processing of camera streams or measurements from an inertial measurement unit (IMU), have to overcome several challenges. First, limited energy supply has to be considered. This particularly happens for battery-based mobile sensing applications such as smartphones and smartwatches. Second, it is necessary to process a potentially big amount of data, which typically arises in applications that use cameras or laser scanners. Third, algorithms need to cope with complicated models that describe the observed process and the measurement system. Finally, algorithms need to be robust to changes in measurement quality. That is, they need to be capable of handling very poor and very precise measurements simultaneously in order to make use of all available information or to account for the fact that some sensors are not available at all time. A typical example for this issue is fusion of information from a magnetometer with other sensors (e.g. a Gyroscope) while accounting for the fact that the former is prone to local disturbances in the magnetic field.

This thesis addresses these challenges by contributing to the field of dynamic state estimation, which is the formal framework for many inference problems including those discussed above. In particular, we are interested in estimation of a not directly observable system state from noisy measurements. This formulation takes both types of uncertainties into account, uncertain system evolution (e.g., wind turbulences affecting aerial vehicles) and uncertainty in measurement systems. Usually, uncertainty is modeled using a set based or a stochastic approach. That is, the uncertainty is either considered to be described by a set of possible values or by a probability distribution. This thesis focuses on the latter case, where system evolution and measurements can be understood as a stochastic process, and thus, the estimation problem can be considered as a stochastic filtering problem. Within the field of stochastic filtering, this thesis addresses the special case of discrete-time filters. This is motivated by the fact that most continuous-time filtering techniques are computationally intractable, and thus, require discretization at some point anyway, e.g., for numerical solution of stochastic differential equations.

We are interested in a very general case of nonlinear discrete-time dynamic state estimation. That is, we assume our system and measurement models to be nonlinear. Furthermore, we consider system states and measurements to be defined not only on linear spaces, but also on nonlinear domains. Particularly, the domains of orientations in Euclidean space (or —more general— the hypersphere) and planar rigid-body motions. In order to obtain tractable algorithms for the considered scenario, we make use of sample-based approaches, which are gaining ever more popularity for dynamic state estimation because they are capable of better capturing the behavior of nonlinear functions. Among these approaches, use of deterministic sampling has emerged as a novel trend in recent years. This is particularly interesting because deterministic sampling schemes offer the advantages of sample-based approaches while simultaneously avoiding the computational burden of random sampling and an absence of deterministic quality guarantees. Thus, we contribute by the development of novel deterministic-sampling techniques and their application to nonlinear dynamic state estimation.

In what follows, a brief overview of state-of-the-art discrete-time filtering techniques will be given. Using three example areas of application, it will be shown where and how improving current filtering algorithms might yield significantly better results. Finally, the contribution of this thesis is outlined followed by an overview of its structure.

1.1 State of The Art

An early discussion of the filtering problem goes back to Kolmogorov [67], [68] and Krein [71], [72] for the discrete-time case and to Wiener [124] for the continuous-time case. However, it was Gauß himself who developed the theory of recursive least squares in [29], [30]. His contribution stayed unnoticed for a long time until over a century later these methods were rediscovered. The first rediscovery happened in the work of Plackett [97] in 1950. The second rediscovery was the seminal work by Kalman [59] in 1960 and Kalman and Bucy [60] in 1961. There, the Kalman filter was proposed for dynamic state estimation in linear systems. Today, it is considered as almost certain that Kalman must have been unaware of the works of Plackett and Gauß because he uses a significantly different derivation. One of the earliest discussions of the Kalman filter from the viewpoint of regression analysis was given in [25].

Application of the ideas of Kalman to nonlinear systems was of considerable interest. Linearizing the system equation using multivariate Taylor series in order to make Kalman filter formulas applicable has been used in the context of the Apollo project [111]. This procedure is known as the extended Kalman filter (EKF). Later, statistical linearization (which was independently developed by Booton [12], [13] and Kazakov [63], [64]) has been used to derive better filters [31], [114] by considering the uncertainty of the current estimate within the linearization procedure. Another approach for handling nonlinear densities was based on approximating the nonlinear density by a mixture of well-known densities. For the Gaussian case, this was done for the first time in [2].

Filters based on random sampling were proposed in order to avoid any type of linearization and to improve the quality of the propagation procedure. Methods based on this approach are known as Particle filters [39] or Sequential Monte Carlo methods [85]. They are discussed more deeply in [4], [23]. Typically, they impose some computational burden for repeated random sampling and suffer from degeneration of particles, which has to be addressed, e.g., by use of resampling techniques.

An entirely deterministic sampling approach was used within the unscented Kalman filter, which was proposed in [54], [55], [53]. It uses $2n + 1$ deterministically placed samples for propagation of n-dimensional uncertain quantities. This contribution can be interpreted as a linear regression Kalman filter [82], [83]. The UKF has sparked off the development of several filtering approaches based on deterministic sampling schemes, because this approaches avoid some of the problems within the particle filter such as the need for performing repeated random sampling of a possibly high number of particles or a potentially poor coverage of the underlying state space. Deterministic sampling was recently also used within advanced filtering techniques, e.g., for introduction of a progressive variant of the measurement update [41] in order to cope with nonlinear measurement models.

Research on nonlinear filtering mostly addressed the problem of handling nonlinear system and measurement functions and did not explicitly consider nonlinear domains. Thus, filters for estimation problems involving a nonlinear underlying domain were mostly based on the Kalman filter or some of its variants discussed above. Therefore, these filters wrongly assumed the underlying domain to be a linear space, which is a good approximation for scenarios with low noise. Examples of these approaches involve [38], [116], [125], [104], [90], [81]. A sound consideration of the underlying domain is made possible by using results from directional statistics [89], which is a subfield of statistics that considers uncertain quantities on nonlinear domains. An approach based on these results was presented in [35]. It used the Bingham distribution for representing uncertain orientations and required random sampling for propagation when the system function was more complicated than a pure rotation. Another thesis [78] that made use of directional statistics was created simultaneously with this work. It mostly focuses on consideration of circular quantities, dependencies between circular quantities, and applications to robotic beating heart surgery.

There is also a huge number of books on stochastic filtering. The following examples might be particularly interesting. Multiple different derivations of the Kalman filter are given in [110]. A discussion of several mostly recent filtering and smoothing techniques can be found in [105]. A fundamental discussion of filtering theory involving both, discrete and continuous time, can be found in [50]. [7], [1], [58]. This brief overview focused mostly on the developments relevant to the contributions made in this thesis. In general, stochastic filtering techniques are a very broad and highly active area of research. A more in-depth historic discussion of stochastic filtering techniques with a stronger emphasis on the case of continuous time can be found in [7].

1.2 Example Areas of Application

The following three examples represent application scenarios that are amongst those that can benefit the most from the development of the contributions made in this thesis. Thus, the challenges involved in these areas are discussed in order to motivate the contributions made in this work.

1.2.1 Mixed and Augmented Reality

Use of a Head Mounted Display (HMD), such as the Oculus Rift, requires tracking of pose and orientation in order to avoid motion sickness. This tracking needs to have a high accuracy for delivering sufficient quality for entertainment applications. Unfortunately, it is insufficient to use an IMU due to possible precision issues and bias. Thus, these systems are typically equipped with an additional external sensor, e.g. a camrea. Therefore, fusion of measurements from both sensors is required in order to produce an accurate estimate. Furthermore, orientation estimation also happens in smartphones, e.g., within gaming and in context of localization applications. In these devices, linear approximations of the underlying domain seem impossible when fusing the magnetometer due to its high uncertainty. Thus, these applications require estimation techniques capable of coping with high uncertainty on nonlinear underlying domains, because assuming local linearity might yield infeasible results.

1.2.2 Intelligent Mobility

One of the early applications of dynamic state estimation techniques within mobility was the use of Global Navigation Satellite Systems (GNSS), such as the Global Positioning System (GPS), the Global Navigation Satellite System (GLONASS), or the Galileo system. In this application, the receiver fuses signals from several satellites in order to obtain a position estimate. Its accuracy can be increased by additionally fusing a correction signal, which is transmitted by a local emitter as it is done in differential GPS. Even if the movement of the underlying object can be approximated by a linear model, this estimation problem still involves the consideration of nonlinearities in the measurement system. Typically, specialized algorithms are used for processing data from a GNSS for position estimation. However, nonlinear filtering techniques are used when additional sensors are involved [107].

A potentially big amount of sensor data needs to be processed by driving assistance systems. For example, modern automatic parking systems carry out several dynamic state estimation procedures. These involve the need to estimate the pose of the car, positions of obstacles for path planning, and tracking of pedestrians who pass by. Further assistance systems, such as lane keeping assistance, traffic jam assistance, or collision avoidance systems typically combine camera-based sensors with radars in order to perform inference about the driving environment. This challenge also exists in driver monitoring systems, such as those estimating the head pose and gaze direction. They are used in order to detect symptoms of fatigue, and thus, potential threats to safety.

The challenges involved in developing driver assistance systems culminate in systems for fully autonomous driving, because estimation algorithms have to cope with less defined environments. This refers to poor road conditions, poor weather conditions, and unpredictable behavior of other drivers and pedestrians. Even though the estimation and decision problems might be similar to the case of driving assistance, a higher robustness is typically required.

Figure 1.1.: Fusing information from intertial sensors in the car and GNSS systems is a typical nonlinear state estimation problem.

1.2.3 Robotic Perception

A major problem in robotics is simultaneous localization and mapping (SLAM) of the environment. Algorithms addressing this problem have made their way into first consumer products, e.g., robotic vacuum cleaners or lawn mowers. These algorithms are strongly related or even entirely based on dynamic state estimation techniques. Examples of filtering approaches in SLAM are algorithms based on the EKF [6] or the UKF [91].

More broadly, dynamic state estimation techniques are used in most areas of robotic perception. This involves sensor calibration [93], sensor fusion [119], and tracking. An early application involving directional statistics was presented for tracking the spin of a ping-pong ball [36].

1.3 Contribution

The examples presented above usually involve some sort of nonlinearity. Either, the system dynamics are described by a nonlinear model or the system state is defined on a nonlinear domain. This thesis contributes techniques that are capable of coping with these nonlinearities by addressing two problems within dynamic state estimation. First, we address the problem of propagation of uncertainty through nonlinear system functions. Second, we address the problem of a sound consideration of underlying domains with a focus on the manifold of uncertain orientations and the manifold of planar rigid-body motions.

1.3.1 Deterministic Sampling of Gaussian Distributions

The main contribution of this thesis is the improvement of existing and the derivation of novel deterministic sampling techniques. That is, we approximate the original continuous distribution by a discrete distribution defined on the same underlying domain. Rather than obtaining this approximation by random sampling, we use entirely deterministic procedures which try to minimize a dissimilarity measure or match some other distributional characteristics. The conceptual differences are visualized in Figure 1.2.

For approximating multivariate Gaussian densities (and Gaussian mixture densities), we use a measure that compares local probability mass at all scales. The contribution of this thesis not only involves investigations of theoretical properties of that measure, but also the derivation of computation formulas for approximating a Gaussian mixture based on numerical optimization or use of precomputed samples. We show that in an important special case of the considered scenario (when both compared distributions have the same mean), the evaluation can actually be performed on all scales, and thus, there is no need to choose a maximum scale as it was done in earlier methods. Furthermore, easily computable formulas are derived for the case of even-dimensional Gaussians.

1.3.2 Domain Specific Probability Distributions

As already mentioned above, algorithms within most real-world applications model uncertainties by assuming a Gaussian distribution or at least a linear underlying state space. An important scenario in which dropping this assumption might yield better results is the handling of uncertain orientations, which can be described by a hyperspherical distribution.

(a) Random Sampling **(b)** Deterministic Sampling

Figure 1.2.: Contrary to random sampling, the deterministic sampling approach that is presented in this thesis offers a better and more homogeneous coverage of the underlying state space.

In particular, unit quaternions can be interpreted as points on the unit hypersphere in $\mathbb{R}^4$. Thus, choosing a suitable probability distribution on this hypersphere makes representation of uncertain orientations possible by representing them as uncertain quaternions. In this work, the Bingham distribution will be used for this representation, because of some convenient properties due to its relation with the Gaussian distribution.

For representation of uncertainties on the manifold of planar rigid-body motions, this thesis proposes a novel distribution reminiscent of the Bingham distribution, and thus, exhibiting similar properties. Its particular advantage is the possibility of a systematically correct consideration of dependencies between position and orientation. Both distributions are employed in order to derive novel filtering techniques for dynamic state estimation on their respective manifolds. Taken altogether, these filters can be thought of as a hyperspherical and $SE(2)$ counterpart to the linear regression Kalman filter (LRKF), and thus, they also require a sampling scheme in order to cope with the manifold-equivalent of a nonlinear system function.

1.3.3 Deterministic Sampling on Nonlinear Domains

Deterministic sampling of uncertain orientations is solved by proposing a more general technique for deterministic sampling on hypershepres of arbitrary dimension. This approach is based on matching the second moment and generating an antipodally symmetric sample set exhibiting this moment. The resulting methodology can be thought of as a hyperspherical analogue to the UKF. Finally, we propose a deterministic sampling scheme for approximating the newly proposed distribution of uncertain planar rigid-body motions. This scheme is based on a hybrid approach. On the one hand, it makes use of the dependency structure within this distribution, and thus, cannot be seen as a purely moment-based approximation. On the other hand, it combines approximations of the Gaussian and the Bingham distributions in order to obtain the final result.

As a result, the newly proposed techniques improve state-of-the-art stochastic filtering algorithms by introducing better approximations of continuous distributions and thus better propagation of uncertainty at low computational cost. Furthermore, they make better consideration of two important nonlinear domains possible by introducing deterministic sampling techniques on these domains and, thus, making stochastic filtering applicable to a broader class of systems.

1.4 Thesis Outline

The entire thesis is subdivided in four parts. In the first part, the considered problem is introduced and motivated (Chapters 1 & 2). In the second part, we consider shape-based deterministic approximation of Gaussian densities (Chapter 3 & 4). Then, deterministic sampling approaches on nonlinear domains are considered (Chapter 5 & 6), which is followed by concluding remarks (Chapter 7). The thesis structure is visualized in Figure 1.3. The following brief overview gives an outlook on the upcoming chapters.

Chapter 2 - Stochastic Filtering

The contributions are motivated by an introductory discussion in Chapter 2. First, modeling of discrete-time dynamical systems and the filtering problem are presented. Second, several techniques for filtering are discussed. This discussion includes the Kalman filter and its adoptions to cope with nonlinearities. Particularly, approaches based on assuming Gaussian uncertainties are considered. They involve well-known filtering techniques based on statistical linearization and moment matching. Additionally, limits of these filtering approaches are presented motivating the need for better algorithms in stochastic filtering. Finally, a formal problem statement describes the problems addressed by this thesis.

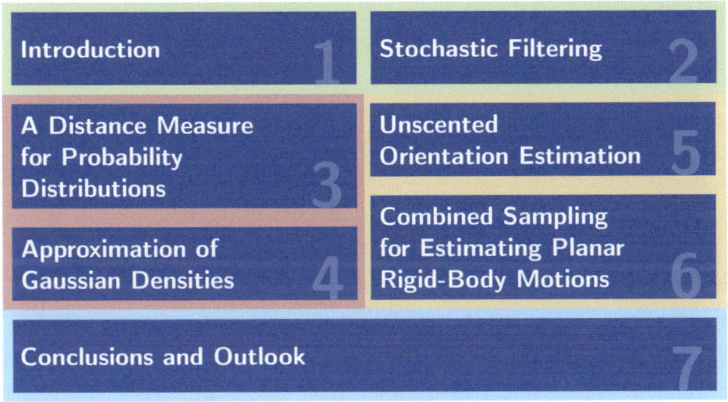

Figure 1.3.: Structure of this thesis.

Chapter 3 - A Distance Measure for Probability Distributions
A distance measure for comparing probability distributions is discussed in Chapter 3. It is based on Localized Cumulative Distributions (LCDs), a modification of the cumulative distribution function that takes local probability mass around a given point into account. It is shown that the Weierstrass transform appears as a special-case in this framework. Furthermore, this chapter contributes by providing a uniqueness result for a special variant of the LCD based on Gaussian kernels. This chapter also discusses some properties of the distance measure and establishes conditions that ensure it to be a metric.

Chapter 4 - Approximation of Gaussian Densities
Subsequently, Chapter 4 contains the main application of the distance measure. That is, a novel approach to approximate multivariate Gaussian distributions (and their mixtures) using a discrete distribution (Dirac mixture) defined on the same domain. This involves overcoming the challenge of multiple integration posed by the computation of the distance measure, which is first discussed for the general case of approximating Gaussian mixtures. Building up on an earlier result, a simplification is proposed for the even-dimensional case of a single Gaussian. First, this is expressing the distance measure in terms of the well-known exponential integral. Second, the distance measure is computed without a limiting

maximum kernel size for the case where both distributions, the Gaussian and the Dirac mixture, have the same mean.

Chapter 5 - Unscented Orientation Estimation

A first consideration of nonlinear domains appears in Chapter 5. There, we propose a purely moment-based deterministic sampling scheme for antipodally symmetric uncertainties on the hypersphere. The proposed scheme is used for approximating the Bingham distribution, which arises naturally when conditioning a Gaussian random vector to unit length. A new filter is derived based on this distribution, which is applicable to orientation estimation based on unit quaternions and generally for hyperspherical estimation problems involving 180° symmetry. Furthermore, a recently proposed saddlepoint approximation is used in order to reduce the computational burden when handling the Bingham normalization constant. The proposed algorithm is compared against a quaternion version of the UKF and the particle filter.

Chapter 6 - Estimation of Planar Rigid-Body Motions

The combination of periodic and non-periodic quantities is considered in Chapter 6. It contains a novel filter for estimation of rigid-body motions in the plane. These are represented by unit dual-quaternions, which are typically used for the more general group of rigid-body motions in Euclidean space. Derivation of the filter is based on two contributions. The first contribution is a novel probability distribution capable of representing uncertain position and orientation simultaneously, and thus, periodic and non-periodic quantities, and dependencies between them. The second contribution is a hybrid deterministic sampling scheme for the newly proposed distribution. It is based on deterministic sampling of both, the Bingham and the Gaussian distributions.

Chapter 7 - Conclusions and Outlook

Finally, the work is concluded in Chapter 7 by summarizing the contributions and discussing interesting open challenges for possible future research.

CHAPTER 2

Stochastic Filtering

Contents

2.1 State-Space Representation

In this work, we consider discrete-time dynamic systems. They are characterized by a system state $\underline{x}_t \in \Omega$ defined on some domain Ω, some control input $\underline{u}_t$, and a transition function a_t (which may be different in each time step). The state evolution is then described by a difference equation

$$\underline{x}_t = a_t(\underline{x}_t, \underline{u}_t) \ .$$

We are interested in the more general case of dynamic systems involving stochastic noise described by $\underline{w}_t$. This system is described by the stochastic difference equation

$$\underline{x}_{t+1} = a_t(\underline{x}_t, \underline{u}_t, \underline{w}_t) \ .$$

For simplicity of presentation, this work will usually not consider any control input $\underline{u}_t$, because this can be thought of as a suitable choice of the transition function. Furthermore, we will only consider time-invariant system dynamics. Nonlinear filters considered in this work can be easily adapted to a time-variant scenario involving control inputs. Thus, the model under consideration here will be given by the stochastic difference equation

$$\underline{x}_t = a(\underline{x}_t, \underline{w}_t) \ .$$

In many applications, it is common to model all arising uncertainties using Gaussian distributions, i.e., in the case above, we would model the noise $\underline{w}_t$ with a Gaussian distribution. In a general setting, this distribution may have a time-dependent mean and covariance matrix. Throughout this thesis, the noise parameters will usually be assumed to be time-invariant. Once again, this purely serves for better presentation. Assuming Gaussian noise for modeling real-world systems is motivated by the central limit theorem (CLT). As there are several central limit theorems, we usually refer to the theorem by Lindenberg & Lévy. In its univariate version, it is formulated as follows.

Theorem 2.1 (Lindenberg-Lévy). *Let x_1, x_2, ... be independent identically distributed (i.i.d.) random variables with mean μ and variance σ^2. Then*

$$\frac{1}{\sqrt{n}} \sum_{i=0}^{n} \frac{x_i - \mu}{\sigma}$$

converges in distribution (i.e., the cumulative distribution function converges pointwise) to a $\mathcal{N}(0, 1)$ distribution as $n \to \infty$.

A proof can be found in most probability theory textbooks, e.g. [108]. Furthermore, there are other versions of the CLT weakening the conditions of this theorem, e.g., allowing for some weak dependence. Thus, it is reasonable to assume that the noise of a real world system follows a Gaussian distribution. There are more reasons for the popularity of this assumption. They are discussed in [65]. It is also typical to use simpler model assumptions in order to ensure tractability of the computations. A first simplification is the assumption of additive noise resulting in the model

$$\underline{x}_{t+1} = a(\underline{x}_t) + \underline{w}_t \ .$$

This is often used when the transition model $a(\cdot)$ is well understood whereas the noise model accounts for all types of uncertainty. Another simplification is the use of an entirely linear model

$$\underline{x}_{t+1} = \mathbf{A}\,\underline{x}_t + \underline{w}_t \ .$$

The use of linear models is highly popular even though most real-world dynamics are nonlinear. This has mainly two reasons. First, it is easier to derive most theory for the linear case. Second, linearization often yields a sufficiently good approximation.

Example 2.2 (Target Tracking). *A typical model in target tracking is the so called constant velocity model [103]. It assumes a target to move in the same direction as in the step before at constant speed. For the 2D case the model is given by*

$$\underline{x}_{t+1} = \begin{pmatrix} 1 & 0 & T & 0 \\ 0 & 1 & 0 & T \\ 0 & 0 & 1 & 0 \\ 0 & 0 & 0 & 1 \end{pmatrix} \underline{x}_t + \underline{w}_t \ ,$$

where $\underline{x}_t$ is a four dimensional state vector and $\underline{w}_t$ a Gaussian noise term. The first two entries of $\underline{x}_t$ describe the position of the target whereas the last two entries describe its speed. T represents the length of the time discretization. Here, the parameters of the noise term $\underline{w}_t$ are typically chosen not only to account for system noise (e.g., turbulences, wind, ...) but also to account for uncertainty in control input of the target, e.g., maneuvers of the target.

Typically, measurement systems are modeled in a similar way. In its full generality, the measurement model can be assumed to be given by

$$\underline{z}_t = h_t(\underline{x}_t, \underline{u}_t, \underline{v}_t) \ .$$

Here $\underline{x}_t$, $\underline{u}_t$ are defined as above, $\underline{v}_t$ accounts for measurement noise, and $\underline{z}_t$ is the actual measurement. The reason for considering a control input in the measurement model is the fact that in some applications, the main goal is the control of measurement systems. A typical example for this type of applications is the sensor scheduling problem.

Example 2.3 (Sensor Scheduling). *Consider a constant velocity system model as given in the previous example. Furthermore, consider a measurement system consisting of two sensors located at $\underline{s}_1$ and $\underline{s}_2$ and performing distance measurements with some distance dependent additive noise. Furthermore, we assume limited energy supply, and thus, only one of the sensor is involved in the typical measurement process. The resulting measurement model can be described by*

$$\underline{z}_t = h(\underline{x}_t, u_t, v_t) = \begin{cases} ||\underline{s}_1 - \underline{x}_t|| \cdot (1 + v_t) & u = 1 \ , \\ ||\underline{s}_2 - \underline{x}_t|| \cdot (1 + v_t) & u = 2 \ . \end{cases}$$

In this scenario, the typical task is choosing an optimal sequence of measurements by a suitable choice of u with respect to some optimality criterion, e.g., uncertainty about the position of the target.

However, all of the discussion on considering a simplified system model is analogously applicable to characterizing uncertain measurement systems. Thus, we will usually assume these to be described by the stochastic difference equation

$$\underline{z}_t = h(\underline{x}_t, \underline{v}_t) \ .$$

2.2 The Stochastic Filtering Problem

Performing inference from uncertain measurements has been the driving factor behind the development of many statistical techniques. Here, we consider the stochastic filtering problem [50], [58], [7]. This problem can be formulated for both, continuous-time and discrete-time scenarios. We consider the discrete-time filtering case. It is characterized by two stochastic processes $\underline{x}_t$ and $\underline{z}_t$ with $t \in \mathbb{N}$ and it is assumed that $\underline{z}_t$ depends on $\underline{x}_t$ and $\underline{x}_t$ is Markovian. The filter problem can be characterized as follows. Given values $\underline{z}_1, \ldots, \underline{z}_t$, we want to obtain an optimal estimate $\hat{\underline{x}}_t$ of $\underline{x}_t$. Optimality is understood in the sense that the estimator minimizes the Mean Square Error (MSE) defined as $\mathbb{E}(||\underline{x}_t - \hat{\underline{x}}_t||^2)$ or in the sense of finding the density $f(\underline{x}_t|\underline{z}_{1:t})$.

The solution to this problem is given by the recursive Bayesian estimator, which is obtained from Bayes' theorem

$$f(\underline{x}_t|\underline{z}_{1:t}) \propto f(\underline{z}_t|\underline{x}_t) f(\underline{x}_t|\underline{z}_{1:t-1}) \ .$$

Computation of the Bayesian estimate might involve several problems. First, obtaining the exact prior $f(\underline{x}_t|\underline{z}_{1:t-1})$ or the exact likelihood $f(\underline{z}_t|\underline{x}_t)$ might be computationally burdensome. Second, the representation of the density $f(\underline{x}_t|\underline{z}_{1:t})$ might become more and more complex over time. Finally, depending on the noise distributions, even simple system functions might result in complicated computations.

In order to apply stochastic filtering for estimating the state of a discrete-time dynamical system as described above, it is necessary to formulate a probabilistic model of the dynamic system. The system evolution can be formulated as

$$f(\underline{x}_{t+1} \,|\, \underline{x}_t) = \int_\Omega \delta(a(\underline{x}_t, \underline{w}_t) - \underline{x}_{t+1}) \, f(\underline{w}_t) \; \mathrm{d}\underline{w}_t \ .$$

Here $\delta(\cdot)$ denotes the Dirac δ-function. In this context it can be interpreted in a measure theoretic sense as a Dirac measure or in a distribution theoretic sense as a Dirac δ-distribution. The measurement likelihood can be formulated as follows

$$f(\underline{z}_t|\underline{x}_t) = \int_\Omega \delta(h(\underline{x}_t, \underline{v}_t) - \underline{z}_t) \, f(\underline{v}_t) \; \mathrm{d}\underline{v}_t \ .$$

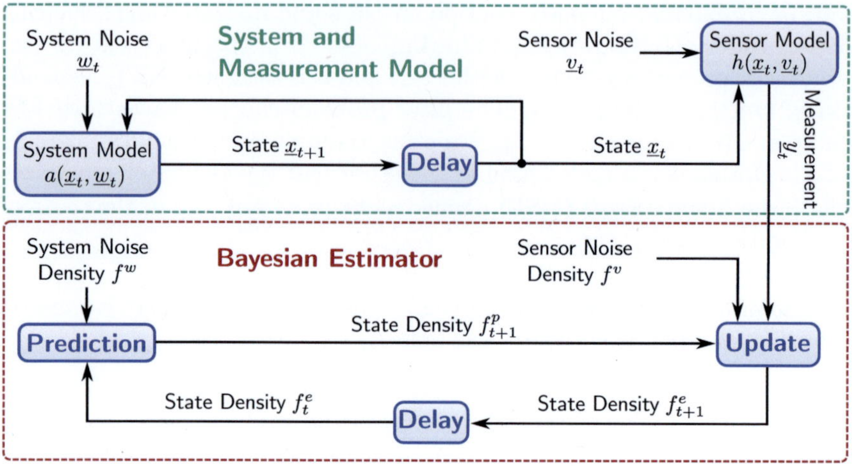

Figure 2.1.: General setup considered in this thesis.

Using this formulation, we can obtain the recursive Bayesian estimator in a two step procedure. In the first step, the system state $\underline{x}_{t+1}$ is predicted using all measurements up to time t by computing

$$\begin{aligned} f(\underline{x}_{t+1} \,|\, \underline{z}_1, \ldots, \underline{z}_t) &= \int_\Omega f(\underline{x}_{t+1}, \underline{x}_t \,|\, \underline{z}_1, \ldots, \underline{z}_t) \, \mathrm{d}\underline{x}_t \\ &= \int_\Omega f(\underline{x}_{t+1} | \underline{x}_t, \underline{z}_1, \ldots \underline{z}_t) \cdot f(\underline{x}_t \,|\, \underline{z}_1, \ldots \underline{z}_t) \, \mathrm{d}\underline{x}_t \\ &= \int_\Omega f(\underline{x}_{t+1} | \underline{x}_t) \cdot f(\underline{x}_t \,|\, \underline{z}_1, \ldots \underline{z}_t) \, \mathrm{d}\underline{x}_t \ . \end{aligned}$$

The second step, known as the measurement update, incorporates the measurement into the estimate by computing

$$\begin{aligned} f(\underline{x}_t | \underline{z}_{1:t}) &= \frac{f(\underline{z}_t \,|\, \underline{x}_t, \underline{z}_{1:t-1}) \cdot f(\underline{x}_t \,|\, \underline{z}_{1:t-1})}{f(\underline{z}_t \,|\, \underline{z}_{1:t-1})} \\ &= \frac{f(\underline{z}_t \,|\, \underline{x}_t) \cdot f(\underline{x}_t \,|\, \underline{z}_{1:t-1})}{f(\underline{z}_t \,|\, \underline{z}_{1:t-1})} \ . \end{aligned}$$

In general, these computations may not be possible in closed form and might be numerically burdensome. Thus, we will discuss a number of filtering techniques dealing with these problems and their limitations. The entire setup outlined in this first two sections is visualized in Figure 2.1.

2.3 Linear Case and the Kalman Filter

As already mentioned, an important special case is the linear case. Here, the entire filter can be carried out in closed form. We assume the system model to be given by

$$\underline{x}_{t+1} = \mathbf{A}\,\underline{x}_t + \underline{w}_t$$

and a linear measurement model

$$\underline{z}_t = \mathbf{H}\,\underline{x}_t + \underline{v}_t \ .$$

Furthermore, we also assume an uncertain initial state characterized by $\mathcal{N}(\hat{\underline{x}}_0, \mathbf{C}_0)$ and all arising noise variables to be Gaussian, independent, and identically distributed (i.i.d.), i.e., $\underline{w}_t \sim \mathcal{N}(\underline{0}, \mathbf{C}_w)$ and $\underline{v}_t \sim \mathcal{N}(\underline{0}, \mathbf{C}_v)$. The prediction step starts with the current estimate $\hat{\underline{x}}_t^e$ and its corresponding uncertainty $\mathbf{C}_t^e$. In order to obtain the estimate of $\underline{x}_{t+1}$ given all information up to time t we compute

$$\begin{aligned}\hat{\underline{x}}_{t+1}^p &= \mathbf{A}\,\hat{\underline{x}}_t^e \\ \mathbf{C}_{t+1}^p &= \mathbf{A}\,\mathbf{C}_t^e\,\mathbf{A}^\top + \mathbf{C}_w\end{aligned}$$

The measurement update is carried out as follows

$$\begin{aligned}\hat{\underline{x}}_t^e &= \hat{\underline{x}}_t^p - \mathbf{K}_t(\underline{z}_t - \mathbf{H}\hat{\underline{x}}_t^p) \ , \\ \mathbf{C}_t^e &= \mathbf{C}_t^p - \mathbf{K}_t\,\mathbf{H}_t\,\mathbf{C}_t^p \ ,\end{aligned}$$

with Kalman gain

$$\mathbf{K}_t = \mathbf{C}_t^p\,\mathbf{H}_t^\top\,\mathbf{S}_t^{-1}$$

and cross-covariance

$$\mathbf{S}_t = (\mathbf{H}_t\,\mathbf{C}_t^p\,\mathbf{H}_t^\top + \mathbf{C}_v) \ .$$

The notation for writing the covariance as $\mathbf{C}$ rather than $\hat{\mathbf{C}}$ is justified because the computation of the covariance matrices does not involve any processing of the measurements and can be carried out in advance. Consequently, the covariance is not an estimate.

This filter is optimal in several respects. It can be shown that the Kalman filter is the best linear estimator even if no Gaussian assumption is made. Furthermore, it is also the optimal Bayesian estimator in the case of Gaussian noise. Consequently, there is also a big number of possible derivations of this filter. The original derivation by Kalman [59] made use of infinite-dimensional Hilbert Spaces. Subsequently, other derivations have been proposed to simplify the presentation. In what follows, we take a Bayesian viewpoint and consider scenarios, where the noise can be assumed to follow a Gaussian distribution.

2.4 Nonlinear Case

Handling of estimation problems in scenarios, where the underlying system is described by a nonlinear model requires the use of approximation techniques. A typical simplifying assumption is assuming the system state and the measurement to follow a jointly Gaussian distribution, and thus, the posterior to be Gaussian. Filters making this assumption are known as Assumed Gaussian Density filters. They are based on purely propagating means and covariances. The goal of this section is to present some of these filters and discuss their respective limitations.

2.4.1 Extended Kalman Filter

The key idea of the extended Kalman filter (EKF) [111] is the linearization of the system and measurement model. The linearization is performed around the current estimate, and thus, carried out according to

$$\mathbf{A}_t = \mathbf{J}_a(\hat{\underline{x}}_t^e) \ , \qquad \mathbf{H}_t = \mathbf{J}_h(\hat{\underline{x}}_t^p) \ .$$

Using these linearized models, the filter equations resemble the original Kalman filter. The prediction step is given by

$$\hat{\underline{x}}_{t+1}^p = a(\hat{\underline{x}}_t^e) \ , \qquad \hat{\mathbf{C}}_{t+1}^p = \mathbf{A}_t \, \hat{\mathbf{C}}_t^e \, \mathbf{A}_t^\top + \mathbf{C}_w \ .$$

And the measurement update is obtained as

$$\begin{aligned}\hat{\underline{x}}_t^e &= \hat{\underline{x}}_t^p - \mathbf{K}_t(\underline{z}_t - \mathbf{H}_t\,\underline{x}_t^p) \ ,\\ \hat{\mathbf{C}}_t^e &= \hat{\mathbf{C}}_t^p - \mathbf{K}_t\,\mathbf{H}_t\,\hat{\mathbf{C}}_t^p \ ,\\ \mathbf{S}_t &= (\mathbf{H}_t\,\hat{\mathbf{C}}_t^p\,\mathbf{H}_t^\top + \mathbf{C}_v) \ ,\\ \mathbf{K}_t &= \hat{\mathbf{C}}_t^p\,\mathbf{H}_t^\top\,\mathbf{S}_t^{-1} \ .\end{aligned}$$

Here, the situation for the covariance matrix differs from the Kalman filter case because it depends on the linearization point, which in turn depends on preceding measurements. Thus, the covariance is an actual estimate and depends on the obtained measurements. The choice of the linearization point is obviously important for the result of the actual estimation process. When the true system state is known, e.g., in case of a deterministic system, linearization yields an accurate local approximation of the system behavior. However, the true system state $\underline{x}_t$ might significantly differ from the estimate $\underline{x}_t^e$, which has a stronger impact on the true system state as the nonlinearity grows.

2.4.2 Statistical Linearization

The problem of choosing the correct linearization point is addressed by statistical linearization. The key idea of filters based on this concept (which were discussed in [31] and under the name of quasi-linear filter in [114]) is to consider the uncertainty of the underlying system state for computing the linear approximate of the true system function. Statistically linearizing a nonlinear function $f(\cdot)$ is carried out by minimizing

$$\arg\min_{\mathbf{F},\underline{b}} \mathbb{E}\left(e(\mathbf{F},\underline{b},\underline{x})^\top \cdot e(\mathbf{F},\underline{b},\underline{x})\right) \ ,$$

where

$$e(\mathbf{F},\underline{b},\underline{x}) := f(\underline{x}) - (\mathbf{F}\cdot\underline{x} + \underline{b}) \ .$$

It can be shown (see [31] eqns. (6.2-7) and (6.2-9)) that this minimization procedure results in

$$\begin{aligned}\mathbf{F} &= \mathbb{E}\left(f(\underline{x})(\underline{x} - \mathbb{E}(\underline{x}))^\top\right)\mathbf{C}_x^{-1} \ ,\\ \underline{b} &= \mathbb{E}(f(\underline{x}) - \mathbf{F}\,\underline{x}) \ .\end{aligned}$$

Here $\mathbf{C}_x$ denotes the covariance matrix of $\underline{x}$. The resulting approximation of $f(\cdot)$ is given by

$$\begin{aligned} f(x) &\approx \mathbf{F}\,\underline{x} + \underline{b} \\ &= \mathbb{E}(f(\underline{x})) + \mathbf{F}\,(\underline{x} - \mathbb{E}(\underline{x}))\ . \end{aligned}$$

This result is reminiscent of the Taylor approximation made by the extended Kalman filter. However, due to consideration of the underlying uncertainty, this approach usually outperforms the extended Kalman filter.

Designing a filter based on this procedure is carried out by statistically linearizing both, the system and the measurement function. Thus, the resulting filter is obtained in the same way as the EKF by replacing the linearized functions in the EKF with their statistically linearized counterparts. For the prediction step, we obtain

$$\begin{aligned} \hat{\underline{x}}^p_{t+1} &= \mathbb{E}(a(\underline{x}^e_t))\ , \\ \hat{\mathbf{C}}^p_{t+1} &= \mathbf{A}_t\,\hat{\mathbf{C}}^e_t\,\mathbf{A}_t^\top + \mathbf{C}_w \\ &= \mathbb{E}\left(a(\underline{x}^e_t)(\underline{x}^e_t - \mathbb{E}(\underline{x}^e_t))^\top\right)(\hat{\mathbf{C}}^e_t)^{-1}\,\mathbb{E}\left(a(\underline{x}^e_t)(\underline{x}^e_t - \mathbb{E}(\underline{x}^e_t))^\top\right)^\top + \mathbf{C}_w\ . \end{aligned}$$

For the measurement update, we first compute

$$\begin{aligned} \underline{v}_t &= \underline{z}_k - \mathbb{E}(h(\underline{x}^p_t))\ , \\ \mathbf{H}_t &= \mathbb{E}\left(h(\hat{\underline{x}}^p_t)(\underline{x}^p_t - \mathbb{E}(\underline{x}^p_t))^\top\right)(\hat{\mathbf{C}}^p_t)^{-1} \end{aligned}$$

and then obtain

$$\begin{aligned} \hat{\underline{x}}^e_t &= \hat{\underline{x}}^p_t - \mathbf{K}\,\underline{v}_t\ , \\ \hat{\mathbf{C}}^e_t &= \mathbf{C}^p_t - \mathbf{K}_t\,\mathbf{H}_t\,\hat{\mathbf{C}}^p_t\ , \\ \mathbf{S}_t &= (\mathbf{H}_t\,\hat{\mathbf{C}}^p_t\,\mathbf{H}_t^\top + \mathbf{C}_v) \\ \mathbf{K}_t &= \hat{\mathbf{C}}^p_t\,\mathbf{H}_t^\top\,\mathbf{S}_t^{-1}\ . \end{aligned}$$

Here and in what follows $\underline{x}^p_t \sim \mathcal{N}(\hat{\underline{x}}^p_t, \hat{\mathbf{C}}^p_t)$ and $\underline{x}^e_t \sim \mathcal{N}(\hat{\underline{x}}^e_t, \hat{\mathbf{C}}^e_t)$. The entire resulting estimator is still an approximate procedure even when the mean is computed analytically. This is not only because of the underlying Gaussian assumption but also because of the approximation made in computing the covariance.

2.4.3 Analytic Approach

The approximate computation of the covariance matrices can be avoided by taking an entirely analytic approach for the moment computation. The prediction step is then obtained by

$$\begin{aligned}
\hat{\underline{x}}_{t+1}^p &= \mathbb{E}(a(\underline{x}_t^e)) \ , \\
\hat{\mathbf{C}}_{t+1}^p &= \mathbb{E}(a(\underline{x}_t^e)a(\underline{x}_t^e)^\top) - \mathbb{E}(a(\underline{x}_t^e))\mathbb{E}(a(\underline{x}_t^e))^\top \ .
\end{aligned}$$

For the measurement update, we obtain

$$\begin{aligned}
\mathbf{S}_t &= \mathbb{E}(h(\underline{x}_t^p)h(\underline{x}_t^p)^\top) - \mathbb{E}(h(\underline{x}_t^p))\mathbb{E}(h(\underline{x}_t^p))^\top \ , \\
\mathbf{K}_t &= \mathbb{E}((\underline{x}_t^p - \mathbb{E}(\underline{x}_t^p))(h(\underline{x}_t^p) - \mathbb{E}(h(\underline{x}_t^p)))^\top)\,\mathbf{S}^{-1} \ , \\
\hat{\underline{x}}_t^e &= \hat{\underline{x}}_t^p + \mathbf{K}_t(\underline{z}_t - \mathbb{E}(h(\underline{x}_t^p))) \ , \\
\hat{\mathbf{C}}_t^e &= \hat{\mathbf{C}}_t^p - \mathbf{K}_t\,\mathbf{S}_t\,\mathbf{K}_t^\top \ .
\end{aligned}$$

From a computational viewpoint, this approach is more burdensome than the statistical linearization approach. The latter can be implemented using $\mathcal{O}(n)$ univariate numerical integration procedures whereas this approach requires $O(n^2)$ univariate numerical integration procedures. However, as both approaches require numerical integration, they might be infeasible for a broad class of real-world applications.

Even though there is no explicit linearization in this approach, it implicitly assumes a linear dependency structure by assuming the state and the measurement to be jointly Gaussian distributed. Thus, this approach is sometimes also referred to as implicit linearization. It is the Gaussian assumption that is the main cause for approximation errors made by the filter. These approximation errors are not only limited to the measurement update step. By assuming the predicted density to be Gaussian, this approach also leads to an approximation error in the prediction step.

2.4.4 Linear Regression Kalman Filter

In order to address the computational burden of numerical integration in both procedures above, an approach based on deterministic sampling has emerged in recent years. Beginning with the unscented Kalman filter (UKF) [53], there has been the development of a great number of methods based on this approach. These approaches are usually known as linear regression Kalman filters [82] or sigma-point filters [118]. They are based on the idea that weighted multivariate linear regression can be applied in order to obtain a linear approximation of the true system function.

Statistical linear regression of a system function as outlined in [82] can be formulated as follows. We consider a set of m deterministic samples $\underline{s}_i \in \mathbb{R}^n$ and a nonlinear function f. We can also assign weights w_i to each of these samples (assuming all $w_i > 0$ and their sum to be 1). These deterministic samples can be interpreted as a probability distribution and they will be used to perform a statistical linear regression of the function f. Let $\underline{s}$ denote a random variable distributed according to this sample distribution. Then, we can define

$$\begin{aligned}
\mathbf{C}_1 &= \mathbb{E}\left(\underline{s}^\top \cdot \underline{s}\right) - \mathbb{E}(\underline{s})^\top \cdot \mathbb{E}(\underline{s}) \ , \\
\mathbf{C}_2 &= \mathbb{E}\left((\underline{s}^\top - \mathbb{E}(\underline{s})) \cdot (f(\underline{s}) - \mathbb{E}(f(\underline{s})))\right) \ , \\
\mathbf{C}_3 &= \mathbb{E}(f(\underline{s})^\top \cdot f(\underline{s})) - \mathbb{E}(f(\underline{s}))^\top \cdot \mathbb{E}(f(\underline{s})) \ .
\end{aligned}$$

Now, statistical linear regression is carried out as follows

$$\arg\min_{\mathbf{F},\underline{b}} \sum_{i=0}^{m} \left(\underline{e}_i(\mathbf{F},\underline{b})^\top \cdot \underline{e}_i(\mathbf{F},\underline{b})\right) \ ,$$

where

$$\underline{e}_i(\mathbf{F},\underline{b}) := f(\underline{s}) - (\mathbf{F} \cdot \underline{s} - \underline{b}) \ .$$

From [3, Theorem 8.2.1] it follows

$$\mathbf{F} = \mathbf{C}_2 \mathbf{C}_1^{-1} \ .$$

This basic idea can now be used to design filters where the estimate of the system state is approximated by a deterministic sample set $\underline{s}_i$ that has the same mean and covariance matrix as the original estimate. The prediction step is then given as follows

$$\begin{aligned}
\hat{\underline{x}}_{t+1}^{p} &= \sum_{i=1}^{m} w_i a(\underline{s}_i) \ , \\
\hat{\mathbf{C}}_{t+1}^{p} &= \sum_{i=1}^{m} w_i \left(a(\underline{s}_i) - \hat{\underline{x}}_{t+1}^{p}\right)\left(a(\underline{s}_i) - \hat{\underline{x}}_{t+1}^{p}\right)^{\top} + \mathbf{C}_w \ .
\end{aligned}$$

For the measurement update, once again, deterministic sampling is carried out. That is, we generate a deterministic sample set $\underline{r}_i$ that has the same mean and covariance matrix as the original predicted state $\hat{\underline{x}}_t^p$. First, this is used to predict the measurement

$$\hat{\underline{z}}_t = \sum_{i=1}^{m} h(\underline{r}_i) \ .$$

Then, we compute the variance of the predicted measurement and their cross covariance.

$$\begin{aligned}
\mathbf{C}_z &= \mathbb{E}\left(h(\underline{r})^{\top} \cdot h(\underline{r})\right) - \hat{\underline{z}}_t^{\top} \cdot \hat{\underline{z}}_t \ , \\
\mathbf{C}_{x,z} &= \mathbb{E}\left((\underline{r}^{\top} - \underline{x}_t^p) \cdot (h(\underline{r}) - \hat{\underline{z}}_t)\right) \ ,
\end{aligned}$$

where $\underline{r}$ is a random variable distributed according to the discrete distribution induced by the deterministic sampling set consisting of the samples $\underline{r}_i$. The remainder of the update is given as follows

$$\begin{aligned}
\mathbf{K}_t &= \mathbf{C}_{x,z}\mathbf{C}_z^{-1} \ , \\
\hat{\underline{x}}_t^e &= \hat{\underline{x}}_t^p + \mathbf{K}_t(\underline{z}_t - \hat{\underline{z}}_t) \ , \\
\mathbf{C}_t^e &= \mathbf{C}_t^p - \mathbf{K}_t \, \mathbf{C}_z \, \mathbf{K}_t^{\top} \ .
\end{aligned}$$

The choice of the sampling scheme is obviously crucial for the quality and performance of this filter, e.g., the UKF matches the first two moments with a fairly small number of samples. A better coverage of the state space can be achieved by combining the concept of deterministic and random sampling or using entirely deterministic approaches, which approximate the shape of the underlying continuous density. These will be discussed in more detail in Chapter 4.

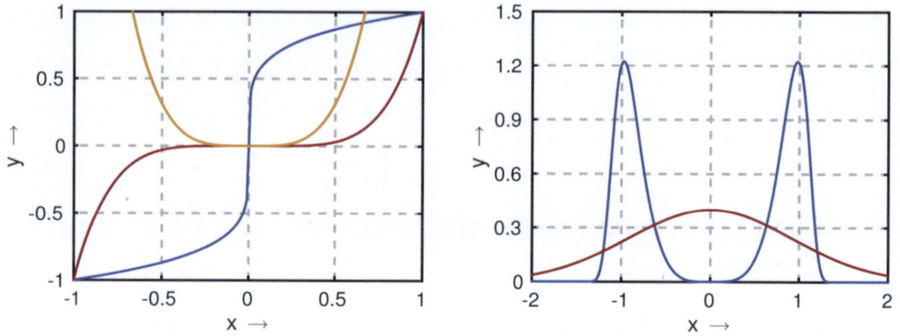

(a) Function through which the Gaussian is propagated (blue), its invert (red), and the derivative of the invert (orange).

(b) True density (blue) and its moment matching based Gaussian approximation (red).

Figure 2.2.: Example where propagation of a single Gaussian through a system-function results in a multimodal distribution. Reapproximating this distribution by a Gaussian might yield an inherently misleading result.

2.4.5 Filtering Based on Gaussian Mixtures

Approximating the distribution of the system state by a Gaussian might be feasible in several scenarios. However, it is very simple to construct an example in which this approximation yields an inherently infeasible result.

Example 2.4. *Consider the model*

$$a(x) = \text{sign}(x) \cdot |x|^{1/5}$$

with

$$\text{sign}(x) = \begin{cases} 1 & x > 0 \,, \\ 0 & x = 0 \,, \\ -1 & x < 0 \end{cases}$$

and let $x \sim \mathcal{N}(0,1)$. *Then, a moment matching based Gaussian approximation of the true distribution of* $a(\underline{x})$ *is obtained by* $\mathcal{N}(\mu, \sigma^2)$ *with* $\mu = \mathbb{E}(a(x))) = 0$ *and* $\sigma^2 = \mathbb{E}(a(x)^2 - \mu^2) \approx 0.841248$. *Furthermore,* $a^{-1}(\underline{x}) = x^5$ *and* $\partial a^{-1}(x)/\partial x = 5\,x^4$.

The true density of $a(x)$ is obtained using the substitution rule, which results in

$$f_a(x) = \left| \frac{\partial a^{-1}(x)}{\partial x} \right| \cdot f(a^{-1}(x))$$
$$= \frac{5x^4}{\sqrt{2\pi}} \exp\left(-\frac{x^{10}}{2}\right) .$$

As can be seen in Figure 2.2, this true density significantly differs from the best Gaussian approximation. Furthermore, this example also shows why the mean is sometimes a very poor estimate of the system-state and why it might be useful to maintain the entire posterior density.

The problem illustrated by this example can be addressed by an approach approximating the true posterior density. This can be done in several ways. An entirely deterministic approach is based on using Gaussian mixtures for filtering [2],[112]. This is justified because of their capability to approximate probability distributions [126]. In their simplest form, Gaussian mixture based approaches simply perform a component-wise extended Kalman filter.

For a filter based on Gaussian mixtures, we assume our estimate to be given by a Gaussian mixture distribution with m components

$$\underline{x}_t^e \sim \sum_{i=1}^{m} p_i \mathcal{N}(\hat{\underline{x}}_{t,i}^e, \mathbf{C}_{t,i}^e) .$$

An early version of a filter that uses Gaussian mixtures (as proposed in [2]) is shown in algorithm 2.1. It is based on the EKF, which is carried out for each mixture component.

This type of filter still requires good handling of nonlinearities due to its componentwise approach. Its performance can be improved by replacing the EKF by one of the more sophisticated filtering techniques discussed above. Furthermore, when the noise also follows a Gaussian mixture distribution, the number of Gaussian mixture components grows exponentially over time. This might require the use of mixture reduction techniques (e.g., [46]), which impose additional computational burden.

Algorithm 2.1 EKF-based Gaussian mixture filter

procedure GMFILTER($\hat{\underline{x}}_{t,1:m}$, $\mathbf{C}^e_{t,1:m}$, $p_{1:m}$, z_{t+1}, $\mathbf{C}_w$, $a(\cdot)$, $\mathbf{C}_v$, $h(\cdot)$)
    for $i \in \{1, \dots, m\}$ **do**
        $\mathbf{A} = \mathbf{J}_a(\hat{\underline{x}}^e_{t,i})$; ▷ Prediction
        $\hat{\underline{x}}^p_{t+1,i} = a(\hat{\underline{x}}^e_{t,i})$;
        $\hat{\mathbf{C}}^p_{t+1,i} = \mathbf{A}\,\hat{\mathbf{C}}^e_{t,i}\,\mathbf{A}^\top + \mathbf{C}_w$;
        $\mathbf{H} = \mathbf{J}_h(\hat{\underline{x}}^p_{t+1,i})$; ▷ Update
        $\mathbf{S} = (\mathbf{H}\,\hat{\mathbf{C}}^p_{t+1,i}\,\mathbf{H}^\top + \mathbf{C}_v)$;
        $p_i \leftarrow p_i \cdot f(\underline{z}_{t+1} - h(\underline{x}^p_{t+1}))$;
        $\hat{\mathbf{C}}^e_{t+1,i} \leftarrow \hat{\mathbf{C}}^p_t - \mathbf{K}_t\,\mathbf{H}_t\,\hat{\mathbf{C}}^p_t$;
        $\underline{x}^e_{t,i} \leftarrow \underline{x}^p_t - \mathbf{K}(\underline{z}_{t+1} - \mathbf{H}_t\,\hat{\underline{x}}^p_{t+1,i})$;
        $\mathbf{K}_t \leftarrow \hat{\mathbf{C}}^p_t\,\mathbf{H}^\top_t\,\mathbf{S}^{-1}_t$; ▷ Reweighting
        $f_i(\underline{x}) \sim \mathcal{N}(\underline{0}, \mathbf{S})$;
    end for
    $c \leftarrow \sum_{i=0}^{m} p_i$; ▷ Renormalization
    for $i \in \{1, \dots, m\}$ **do**
        $p_i \leftarrow p_i / c$;
    end for
    return $(\hat{\underline{x}}_{t+1,1:m}, \mathbf{C}^e_{t+1,1:m}, p_{1:m_t})$;
end procedure

Entirely different problems may arise when the state space is a nonlinear domain because assuming a distribution defined on a linear domain is inherently wrong in these cases and not necessarily a good approximation. The main reason for this is the fact that usually these approximations not only involve false assumptions about the true underlying distribution, but also make false assumptions about the geometry of the underlying domain. Thus, it is important to know in which cases these approximations are justified.

2.5 Problem Formalization

The problems discussed so far motivate the contribution of this thesis. In order to cope with nonlinearities, it is necessary to be capable of approximation probability densities efficiently and accurate. Furthermore, it is necessary to correctly consider a nonlinear underlying domain.

2.5.1 Problem 1: Density Approximation

For improving performance of deterministic sampling based filters our first goal is the approximation of Gaussian densities (or mixtures of Gaussian densities) by a discrete density defined on the same domain. That is, for a Gaussian (mixture) described by its density $f_G(\cdot)$, we want to obtain a discrete distribution described by

$$f_D(\underline{x}) = \sum_{i=1}^{m} w_i \delta(\underline{s}_i - \underline{x})$$

in a way minimizing some dissimilarity measure $D(f_G, f_D)$. This is motivated by the need for computing expectations of the type $\mathbb{E}(g(\underline{x}))$, where $\underline{x}$ is a Gaussian (mixture) random vector. This can be directly used in a linear regression Kalman filter or as a propagation scheme in a Gaussian mixture based filter.

On the one hand, use of very rough approximations might be computationally tractable but yield poor results. On the other hand, use of adaptive numerical integration techniques might result in very accurate computation of the expectation. However, as mentioned above, this might be computationally intractable. The same problem might occur for Monte Carlo integration requiring a repeated-random sampling procedure. Thus, the main objective in the development of our deterministic sampling procedure is to offer a good tradeoff between inaccuracy and computational burden. Due to the widespread use of Gaussian distributions, this result is of particular importance for the Gaussian special case.

2.5.2 Problem 2: Consideration of Nonlinear Domains

For stochastic filtering on nonlinear domains, it is in general not sufficient to simply adapt the sampling scheme. A correct description of uncertainties on these domains is required. Among all nonlinear domains, the unit hypersphere $S^{n-1} \subset \mathbb{R}^n$ and the manifold of planar rigid-body motions are of particular interest. For the hyperspherical case, we are particularly interested in the special case $n = 4$, because a 4-dimensional unit hypersphere can be used to represent orientations in 3d space.

Thus, our goal is to discuss or even propose new probability distributions, which can be used to represent uncertainty on the hypersphere. Furthermore, we want to derive deterministic approximation schemes on these domains. Finally, the challenge lies in using these results to derive new filtering techniques on these domains.

CHAPTER 3

A Distance Measure for Probability Distributions

Contents

3.1 Overview

Measures for comparing two probability distributions are motivated by applications such as parameter estimation, hypothesis testing, data compression, or density approximation. These measures are known as *probability metrics* in literature [98], [127]. Furthermore,there are several measures motivated by information theory, such as the Kullback-Leibler divergence [74] or the —more general— Rényi divergence [101]. Most of these measures usually suffer from one of the following drawbacks. For some measures, it is not possible to evaluate the measure in a computationally tractable way (e.g., because they are formulated as an infimum over all joint distributions between the two originally compared distributions.). In these situations, bounds are typically used to approximate the true value of the measure [32]. Furthermore, some similarity measures are not applicable in context with discrete probability distributions or require that at least one of the compared distributions is absolute continuous (in the measure-theoretic sense) with respect to the other. This makes comparison of continuous and discrete probability distributions impossible. Finally, some measures are not symmetric, and thus, their interpretability is difficult for several applications.

In this chapter, we investigate the properties of a similarity measure that addresses several of the problems outlined above. Its key idea is to perform a transformation of a probability distribution in order to achieve a representation as a continuous function that can be used as a basis for a L_2 distance measure. The transform is based on a convolution transform with a suitable kernel function that characterizes the area of considered probability mass, which is typically chosen in a way putting stronger emphasis around a given point. These so-called Localized Cumulative Distributions (LCDs) can be thought of as a generalization of the cumulative distribution function (CDFs). This entire framework was originally proposed in [44] and further elaborated on in [43]. Here, we revisit these definitions and prove new results providing a theoretical justification for considering this representation of probability distributions. In an important special case, we show that the LCD is a unique representation of a probability distribution and that the proposed similarity measure is a metric. Thus, a number of desirable properties, e.g. symmetry or positive definiteness, are maintained. Furthermore, a relationship to the generalized multivariate Weierstrass transform is used for the derivation of an inversion formula.

In what follows, we will first discuss some existing similarity measures and their respective problems. This is followed by an introduction to LCDs and the corresponding distance measure. In this general framework, we then consider the special case of the LCD transform that is based on Gaussian kernels and discuss its relationship with the Weierstrass transform. The chapter is concluded by a discussion of possible application and some summarizing remarks.

3.2 Existing Similarity Measures

One of the most widely used approaches for comparing probability distributions is the Kullback-Leibler divergence (often also referred to as information gain or relative entropy). It is defined as

$$d_{KL}(P_1, P_2) = \int_{\Omega} f_1(\underline{x}) \ln\left(\frac{f_1(\underline{x})}{f_2(\underline{x})}\right) \mathrm{d}\underline{x} \ ,$$

with f_i being the density of P_i. It has a natural information theoretic interpretation as the amount of information lost when P_1 is approximated by P_2. The definition above can be adapted to discrete distributions defined on the same set by replacing the densities f_i by the corresponding probability mass functions. However, this relationship is not symmetric, and thus, neither is the measure itself. Even though the Kullback-Leibler divergence is defined for both, continuous and discrete probability distributions, it is incapable of comparing both types simultaneously. These problems also exist in the Rényi divergence.

A true metric for probability distributions on some metric space was proposed by Wasserstein [123]. In its general form, it is defined as

$$W_p(P_1, P_2) = \inf_{P \in \mathcal{J}(P_1, P_2)} \left(\mathbb{E}_P(d(\underline{x}, \underline{y})^p)\right)^{1/p} \ ,$$

where $\underline{x} \sim P_1$, $\underline{y} \sim P_2$, and $\mathcal{J}(P_1, P_2)$ is the set of all joint probability distributions with marginals P_1 and P_2. An intuitive interpretation of this metric is related to transportation theory. That is, the metric can be viewed as the minimum cost of transforming one probability distribution into another. It does not only obey the axioms of a metric but is also capable of handling continuous and discrete distributions simultaneously. There are several applications of this metric in certain special cases such as [10]. Unfortunately, in the general case, computation of this metric becomes intractable, because the infimum is taken over all possible joint distributions.

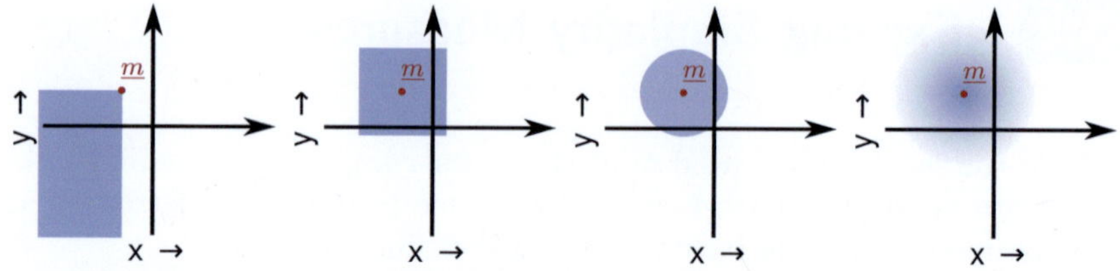

(a) CDF **(b)** Rectangular **(c)** Unit Disk **(d)** Gaussian RBF

Figure 3.1.: Different possible kernel choices for Localized Cumulative Distribution. This results in a generalization of several concepts of representing a probability distribution.

A computationally feasible approach for comparing the shape of continuous probability densities is based on the L^2 norm

$$||f_1 - f_2||_{L^2} = \sqrt{\int_{\Omega} (f_1(\underline{x}) - f_2(\underline{x}))^2 \, \mathrm{d}\underline{x}} \ .$$

This also has the desirable property of being an actual metric and thus, satisfying intuitively desirable properties such as symmetry and the triangle inequality. Replacing densities by probability mass functions and the integral by a sum makes this measure applicable to comparing two discrete distributions. Unfortunately, similarly to the Kullback-Leibler divergence, it is also not capable of comparing a discrete with a continuous probability distribution.

3.3 Localized Cumulative Distributions

In the following, we define a very general transform of probability distributions that makes the intuitive idea of representing local probability mass precise. This is achieved by taking expectation over a kernel function that characterizes the area of considered probability mass around a given point. The following definition is a generalized version of the original (as proposed in [44]). In its general formulation it was originally proposed in [43].

Definition 3.1. *Consider a probability distribution P and a random variable $\underline{x} \sim P$. Then the Localized Cumulative Distribution (LCD) of P is given by*

$$\mathcal{L}_K[P](\underline{m}, b) := \mathbb{E}\left(K(\underline{x} - \underline{m}, \underline{b})\right) \ ,$$

where $K(\cdot, \cdot)$ is a suitable kernel, which is located at $\underline{m}$ and characterized by $\underline{b}$ (which will be usually used to represent the kernel width). We call $K(\underline{x}, \underline{b})$ separable, if it can be decomposed according to

$$K(\underline{x}, \underline{b}) = \prod_{i=1}^{n} K(\underline{x}^{(i)}, \underline{b}^{(i)}) \ .$$

For brevity, we will write $\mathcal{L}[P](\underline{m}, \underline{b})$ whenever the chosen kernel is clear from the context. The notation $\mathcal{L}[f](\underline{m}, \underline{b})$ will be used for the LCD transform of the probability density f of P.

Typical representations of probability distributions such as the CDF can now be seen as a special case of the LCD. Unfortunately, the existence of an LCD representation is not guaranteed and highly dependent on the combination of distribution and kernel function. Thus, the choice of the kernel is of particular importance. Several examples for kernel functions for a 2d distribution are shown in Figure 3.1. In order to account for the basic idea of representation of localized probability mass, we focus on kernels symmetric around $\underline{m}$.

The LCD representation has several desirable properties. In [44], it was pointed out that the LCD of $\underline{x}$ can be decomposed into the LCD of all entries x_i if the kernel K is separable (in the sense that it can be written as a product of kernels K_i each of which only affects x_i) and all entries are pairwise independent, that is

$$F(\underline{m}, \underline{b}) = \prod_{i=1}^{n} F_i(m_i, b_i) \ ,$$

where $F_i(\cdot, \cdot)$ denotes the LCD of $\underline{x}_i$.

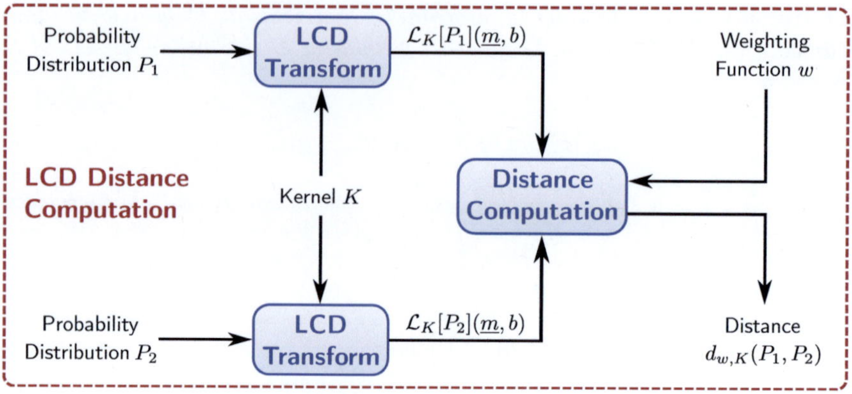

Figure 3.2.: LCD distance computation scheme.

3.3.1 A Distance Measure for Probability Distributions

The definition of the LCD was motivated by the need for comparing probability distributions, particularly by the need for comparing discrete and continuous distributions simultaneously. The key idea for the definition of the distance measure is to compare the entire state space and to compare probability masses on all scales. That is, the distance measure is based on comparing the LCD representations of two probability distributions at all kernel positions and for all possible kernel sizes. These considerations result in the following definition.

Definition 3.2. *Consider two probability distributions P_1, P_2 defined on Ω. Then, the LCD distance between P_1 and P_2 is*

$$d_{w,K}(P_1, P_2) := \sqrt{\int_0^\infty w(b) \int_\Omega (\mathcal{L}_K[P_1](\underline{m}, b) - \mathcal{L}_K[P_2](\underline{m}, b))^2 \, \mathrm{d}\underline{m} \, \mathrm{d}b} \, ,$$

where $K(\cdot, \cdot)$ is the LCD Kernel and $w(\cdot)$ a positive weighting function.

The entire computation process of this distance is visualized in Figure 3.2. Whether this definition is meaningful is highly dependent on the combination of the considered weighting function w and LCD kernel K. At the very least, K and w need to be chosen properly to ensure that the integrals defining the distance converge. In order for $d_{w,K}$ to be a metric, it needs to satisfy the usual conditions

$$\begin{aligned}
&\text{(M1) } d_{w,K}(P_1, P_2) \geq 0 \ , \\
&\text{(M2) } d_{w,K}(P_1, P_2) = 0 \Leftrightarrow P_1 = P_2 \ , \\
&\text{(M3) } d_{w,K}(P_1, P_2) = d_{w,K}(P_2, P_1) \ , \\
&\text{(M4) } d_{w,K}(P_1, P_2) \leq d_{w,K}(P_1, P_3) + d_{w,K}(P_3, P_2) \ .
\end{aligned}$$

The conditions (M1) and (M3) hold as a consequence of our definition of the LCD distance. For (M2) to hold, it is necessary that the LCD representation is unique and $w(b)$ is not vanishing for at least some interval where $\mathcal{L}_K[P_1](\underline{m}, b) \neq \mathcal{L}_K[P_2](\underline{m}, b)$. The triangle inequality (M4) is shown using the typical proof with the help of the Cauchy-Schwarz-Bunyakowsky (CSB) inequality. For simplicity, let $\underline{s} := (b, \underline{m}^\top)^\top$, $M := \mathbb{R}^+ \times \Omega$, and

$$\begin{aligned}
A_1(\underline{s}) &:= \sqrt{w(b)} \left(\mathcal{L}_K[P_1](\underline{m}, b) - \mathcal{L}_K[P_3](\underline{m}, b) \right) \ , \\
A_2(\underline{s}) &:= \sqrt{w(b)} \left(\mathcal{L}_K[P_3](\underline{m}, b) - \mathcal{L}_K[P_2](\underline{m}, b) \right) \ .
\end{aligned}$$

Then, we obtain

$$\begin{aligned}
d_{w,K}(P_1, P_2)^2 &= \int_M \left(A_1(\underline{s}) + A_2(\underline{s}) \right)^2 \, \mathrm{d}\underline{s} \\
&= \int_M A_1(\underline{s})^2 + 2\, A_1(\underline{s})\, A_2(\underline{s}) + A_2(\underline{s})^2 \, \mathrm{d}\underline{s} \\
&\leq \int_M A_1(\underline{s})^2 \, \mathrm{d}\underline{s} + 2 \left| \int_M A_1(\underline{s})\, A_2(\underline{s}) \, \mathrm{d}\underline{s} \right| + \int_M A_2(\underline{s})^2 \, \mathrm{d}\underline{s} \\
&\overset{\text{CSB}}{\leq} \int_M A_1(\underline{s})^2 \, \mathrm{d}\underline{s} + 2 \sqrt{\int_M A_1(\underline{s})^2 \, \mathrm{d}\underline{s}} \cdot \sqrt{\int_M A_2(\underline{s})^2 \, \mathrm{d}\underline{s}} \\
&\qquad + \int_M A_2(\underline{s})^2 \, \mathrm{d}\underline{s} \\
&= \left(\sqrt{\int_M A_1(\underline{s})^2 \, \mathrm{d}\underline{s}} + \sqrt{\int_M A_2(\underline{s})^2 \, \mathrm{d}\underline{s}} \right)^2 \\
&= \left(d_{w,K}(P_1, P_3) + d_{w,K}(P_3, P_2) \right)^2 \ .
\end{aligned}$$

Finally, (M4) results as a consequence of the monotonicity of the square root.

When the Gaussian kernel is used, possible choices for the weighting factor $w(b)$ might be

$$w(b) = \begin{cases} b^{-n+1} , & b \in [0, b_{max}) \\ 0 , & \text{otherwise} \end{cases}$$

or even $w(b) = b^{-n}$. The distance measure presented here can be intuitively interpreted as describing the similarity between the shapes of two probability distributions. Comparison on all scales b guarantees, that not only local dissimilarities are considered. Finally, the proposed weighting function guarantees that a stronger emphasis is put on local dissimilarities in characterizing the distance.

3.4 Gaussian Kernels

Particularly, the Gaussian radial basis function (RBF), will be used in the further work. It is given by

$$K(\underline{x}, b) = \exp\left(-\frac{||\underline{x} - \underline{m}||_2^2}{2b^2} \right) .$$

With this choice, we have existence of the LCD of an arbitrary probability distribution. Furthermore, using RBFs guarantees uniqueness of the LCD as a representation of the original distribution. This is established by the following result.

Theorem 3.3 (Uniqueness). *Consider the probability distributions P_1, P_2 and their corresponding LCDs $F_1(\underline{m}, b)$, $F_2(\underline{m}, b)$. Then $F_1(\underline{m}, b) \equiv F_2(\underline{m}, b)$ if and only if $P_1 = P_2$.*

Proof. If $P_1 = P_2$ then the equality of the LCD representation follows from the definition of the LCD. Thus, we only need to consider the case where $F_1(\underline{m}, b) \equiv F_2(\underline{m}, b)$. Let $\underline{x}_1 \sim P_1$ and $\underline{x}_2 \sim P_2$. Furthermore let $c(b)$ denote the normalization constant of a $\mathcal{N}(\underline{0}, b^2\,\mathbf{I})$ distribution, that is

$$c(b) = \frac{1}{(2\pi)^{n/2}\, b^n} .$$

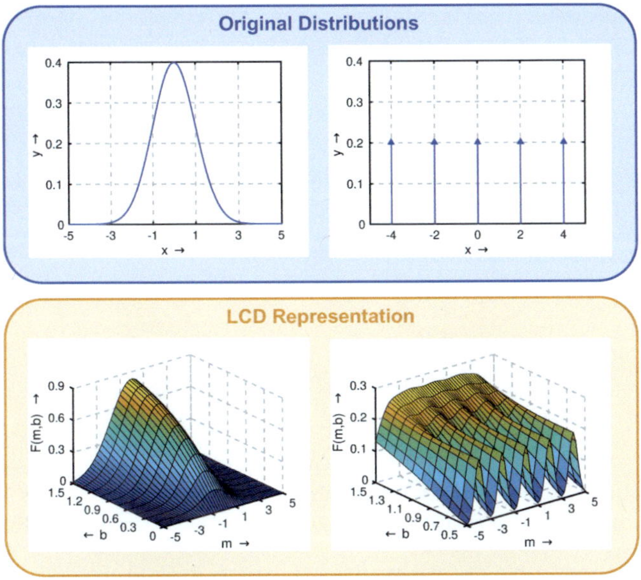

Figure 3.3.: Probability density function of a standard normal Gaussian and probability mass of a discrete distribution and their corresponding representations as LCD.

Furthermore, let $\underline{z} \sim \mathcal{N}(\underline{0}, \mathbf{I})$. Then the density of $\underline{y}_{b,1} := \underline{x}_1 + b\,\underline{z}$ is given by $f_1(\underline{x}) = c(b)\,F_1(\underline{x}, b)$. The density $f_2(\underline{x})$ of $\underline{y}_{b,2} := \underline{x}_2 + b\,\underline{z}$ is defined analogously and we have $f_1(\underline{x}) = f_2(\underline{x})$. Moreover, we also obtain almost sure convergence of $\underline{y}_{b,1}$ to $\underline{x}_1$ by

$$\mathbf{P}\left(\lim_{b \to 0} \underline{y}_{b,1} = \underline{x}_1\right) = \mathbf{P}\left(\lim_{b \to 0} (\underline{x}_1 + b\underline{z}) = \underline{x}_1\right) = 1$$

and an analogous result for almost sure convergence of $\underline{y}_{b,2}$ to $\underline{x}_2$. Almost sure convergence implies convergence in distribution (see Theorem 9.2.1 and Proposition 9.3.5 in [24]). Thus, we obtain

$$\underline{x}_1 \overset{d}{=} \lim_{b \to 0} \underline{y}_{b,1} \overset{d}{=} \lim_{b \to 0} \underline{y}_{b,2} \overset{d}{=} \underline{x}_2 \ ,$$

where " $\overset{d}{=}$ " denotes equality in distribution. Consequently, $P_1 = P_2$. □

Using the Gaussian kernel results in a smooth representation of probability distributions even if the original distribution has been discrete (see example in Figure 3.3).

Example 3.4. *LCD Computation for different distribution types.*

a) The LCD of a standard normal distribution is obtained as

$$\begin{aligned}\mathcal{L}_K[\mathcal{N}(0,1)](\underline{c},b) =& \mathbb{E}(K(x-c,b)) \\ =& \int_{\mathbb{R}} \exp\left(-\frac{(x-c)^2}{2b^2}\right) \frac{1}{\sqrt{2\pi}} \exp\left(-\frac{x^2}{2}\right) \, \mathrm{d}x \\ =& \frac{1}{\sqrt{1+b^2}} \exp\left(-\frac{c^2}{2(1+b^2)}\right)\end{aligned}$$

b) The LCD of discrete distributions is somewhat easier. For a one-dimensional probability distribution P defined on the points $-4, 2, 0, 2, 4$ with equal probability, the LCD is a weighted sum of the kernels, that is

$$\begin{aligned}\mathcal{L}_K[P](c,b) =& \mathbb{E}(K(x-c,b)) \\ =& \frac{1}{5}\sum_{i=1}^{5} K(s_i - c, b) \ ,\end{aligned}$$

where $x \sim P$.

Both examples are visualized in Figure 3.3.

3.4.1 Relationship to Weierstrass Transform

The Gaussian LCD of a univariate probability distribution with density f is related to the Weierstrass transform of a function f, which is defined as

$$\mathcal{W}[f](c) := \int_{-\infty}^{\infty} \frac{1}{\sqrt{4\pi}} \exp\left(-\frac{c-x}{4}\right) f(x) \, \mathrm{d}x$$

or —in its generalized variant— as

$$\mathcal{W}_a[f](c) := \int_{-\infty}^{\infty} \frac{1}{\sqrt{4\pi a}} \exp\left(-\frac{c-x}{4a}\right) f(x) \, \mathrm{d}x \ .$$

Thus, we have

$$\mathcal{L}[f](m, b) = \sqrt{2\pi b^2} \cdot \mathcal{W}_{b^2/2}[f](m) \ .$$

This relationship gives rise to a number of interesting results already available for the Weierstrass transform, e.g., the existence of an inversion formula

$$\mathcal{W}^{-1}[\mathcal{W}[f]](c) = \sum_{n=0}^{\infty} v_n H_n(x/2) \ ,$$

where v_n are coefficients of a series expansion of $\mathcal{W}[f]$ (assuming it to be analytic). Thus, an inversion formula for the LCD representation can be given by

$$f(c) = \mathcal{W}^{-1}\left[\frac{\mathcal{L}[f](\cdot, \sqrt{2})}{\sqrt{2\pi b^2}}\right](c) \ .$$

A generalization to the multivariate case requires the use of the multivariate Weierstrass transform and further development of its theory. Its application in the context of probability theory has been discussed in [122].

3.5 Summary and Discussion

In this chapter, we revisited a smooth representation of probability distributions called the localized cumulative distribution. It is based on a convolution transform with a properly chosen kernel function. Furthermore, a distance measure based on this representation was considered. We have shown that this distance measure is a true metric in several special cases. One of these cases is choosing Gaussian kernel functions. We have shown this choice to yield an LCD uniquely representing a probability distribution. Furthermore, a relationship to the Weierstrass transform made results involving this transform easily applicable to the LCD, e.g., the derivation of an inversion formula.

The considered density representation and the corresponding distance can be utilized in a wide range of applications, whenever a comparison of probability distributions is necessary. These applications involve the approximation of a discrete probability distribution on a continuous domain by another distribution with a smaller number of Dirac-Delta components [42], parameter estimation [70], and nonlinear stochastic model predictive control [19]. In the next chapter, we will discuss one particular application in more detail. The representation discussed here will be used for an entirely deterministic approximation procedure for approximation of Gaussian densities and Gaussian mixtures. Areas of possible future research involve application of the proposed measure to hypothesis testing, information theoretic control, and a deeper investigation of its theoretical properties.

CHAPTER 4

Approximation of Gaussian Densities

Contents

4.1 Introduction

In this chapter, we consider the approximation of Gaussian distributions (including the more general case of Gaussian mixtures) by a discrete distribution defined on the same continuous domain. Our approach is based on using a global comparison measure which was presented in the previous chapter. This can be thought of as finding a Dirac mixture that is most similar (in shape) to the given Gaussian (mixture).

First, we will introduce a very general approximation scheme for Gaussian mixtures. This is motivated by the fact that a convex combination of Gaussian densities can be used to approximate other probability densities [126]. The main challenge is the derivation of a computationally tractable approximation scheme. Particularly, it is important to simplify the computation of the LCD distance measure. Use of a naïve entirely numerical evaluation of this distance is intractable, because it would require the computation of a $(n+1)$-dimensional integral when performing a comparison of two n-dimensional distributions. Here, we contribute by removing the need for multiple numerical integration leaving one integral which can be evaluated numerically. This generalizes a result from [43], where this was done for one single Gaussian.

Second, we consider the special case of a Gaussian density. Approximations of these densities are not only important as a special case of the Gaussian mixture, but also motivated by the fact that the Gaussian distribution arises as the limit distribution in the central limit theorem. Its consideration is more than merely a special case of the Gaussian mixture, because in the case of approximating standard normal even-dimensional densities several important simplifications occur making the formulation of the distance measure in terms of well-known mathematical functions possible. This narrow case can be generalized to a suboptimal (w.r.t. the distance measure) sampling scheme of arbitrary even-dimensional Gaussians.

Finally, in the case where the mean of the Gaussian and the mean of the Dirac mixture are equal, we can obtain another simplification in the computation. This is based on removing the need for choosing integration bounds by cutting off the weighting function within the distance measure. This does not only result in simpler practical applications, but also simplifies the actual computation of the distance measure.

The proposed density approximation approach is evaluated in simulations by comparing it to other density approximation approaches. We evaluate both, the optimal online approximation and the faster approach based on precomputing an approximation of the standard Gaussian and then applying a suitable transformation in order to obtain the desired mean and covariance. We show that both, online and offline sampling approaches outperform comparable state-of-the-art approximation techniques.

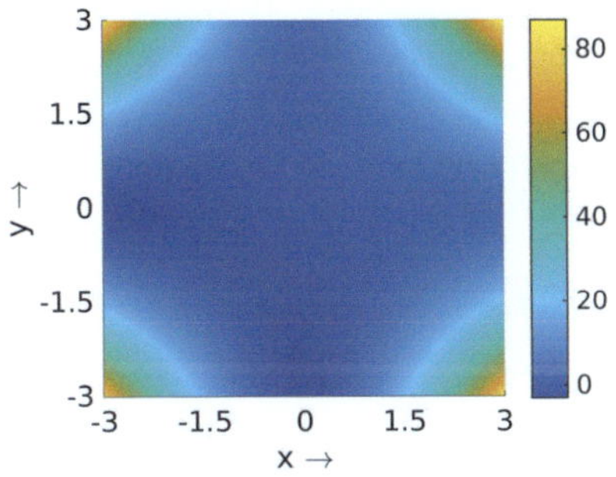

Figure 4.1.: Example function for which approximate propagation by placing samples on the main axes yields wrong results.

The results in this chapter are partially based on and extend those presented in [113], [135], [141]. A filter based on deterministic sampling procedures that minimize the LCD distance was presented in [113].

4.2 Related Approaches

A naïve distribution approximation approach is the use of random sampling as is done in particle filters [23]. Propagation of uncertainty using these approaches is shown to converge (as the number of samples grows) in many cases. However, there is a number of drawbacks. First, a higher number of samples is required compared to deterministic positioning. Second, the process of random sampling might be computationally burdensome. Finally, reproducibility of results can not be maintained unless a predefined seed is used for generation of pseudo-random numbers.

The unscented transform [53], [54], [55] presents an approximation of the first two moments requiring $2n + 1$ samples in n dimensions. It places the samples along the principal axes of the described covariance. In a later work, a version requiring $n + 2$ samples was proposed [52]. A filter generalizing the original idea of the unscented transform was presented in [47]. The algorithm proposed therein accounts for the fact that $2n + 1$ samples might be insufficient for capturing the behavior of the considered system function. Thus, it offers the possibility to place an arbitrary predefined number of samples on the main axes of the covariance ellipsoids which yields better results. This, in turn, might be insufficient for propagating uncertainties through functions where the interesting behavior happens outside the considered axes as will be seen in the following example.

Example 4.1. *Consider the propagation of a standard normal bivariate random variable* $\underline{w} \sim \mathcal{N}(\underline{0}, \mathbf{I})$ *through a function* $g : \mathbb{R}^2 \to \mathbb{R}$ *(shown in Figure 4.1) given by*

$$g(\underline{x}) = x^{(1)} + x^{(2)} + (x^{(1)} \cdot x^{(2)})^2 \ .$$

Let $\underline{\tilde{w}}_n$ *denote a random variable whose distribution is meant to approximate* $\mathcal{N}(\underline{0}, \mathbf{I})$ *by placing values symmetrically on the main axes while maintaining the first two moments. Then,* $\mathbb{E}(g(\underline{\tilde{w}}_n)) = 0$ *for all* n *whereas the true expectation is*

$$\mathbb{E}(g(\underline{w})) = \underbrace{\mathbb{E}(w^{(1)})}_{=0} + \underbrace{\mathbb{E}(w^{(2)})}_{=0} + \underbrace{\mathbb{E}((w^{(1)})^2)}_{=1} \cdot \underbrace{\mathbb{E}((w^{(2)})^2)}_{=1} = 1 \ .$$

Thus, approximation approaches with an arbitrary number of samples might fail to converge as long as all of these samples are placed on the main axes of the covariance ellipsoid.

The problem outlined in the preceding example is addressed in a further generalization of the unscented transform, which is discussed in [26], [115]. The basic idea of this approach is using several UKF sample sets simultaneously. Each of these sample sets is rotated with a random angle, and the spread of the samples is also chosen randomly. Thus, the resulting approach combines deterministic and random sampling and we would expect it to outperform randomized approaches and yield worse results compared to entirely deterministic approximation schemes covering regions with high probability mass.

The approach discussed in context of the Gaussian Hermite Kalman filter (GHKF) proposed in [48] is based on using a grid. That is, the approximation procedure comes down to choosing the respective probability weights for each grid point. This approach is particularly prone to the curse of dimensionality, because the number of grid points always grows exponentially with the dimension, even if the approximated density has almost vanishing variance in all but a few of the main axes of its covariance ellipsoid. Some sampling approaches that were discussed here are visualized in Figure 4.2 and an overview is given in Table 4.1.

Approach	**Sample Count**	**Method**	**References**
Gauss Filter	$L \cdot n$	deterministic	[47]
Gauss-Hermite KF	L^n	deterministic	[48]
LCD	L	deterministic	[43], [113], [135], [141], this thesis
Particle Filter	L	random	[4], [23],[39]
Randomized UKF	$2 \cdot L \cdot n + 1$	mixed	[26], [115]
Spherical Simplex UKF	$n + 2$	deterministic	[52]
UKF	$2 \cdot n + 1$	deterministic	[53], [54], [55]

Table 4.1.: Overview of different sampling approaches used in stochastic filters. In the number of samples n denotes the dimension and $L \in \mathbb{N}^+$ is a constant that may be chosen by the user.

4.3 Formal Problem Statement

We consider a Gaussian mixture distribution in $\mathbb{R}^n$ given by its density

$$f_{GM}(\underline{x}) = \sum_{i=0}^{l} p_i \, f_i(\underline{x})$$

where f_i is the density of a Gaussian $\mathcal{N}(\underline{\mu}_i, \mathbf{\Sigma}_i)$ distribution. Given a predefined number of samples m and their respective weights w_i $(i = 1, ..., m)$, the first goal considered in this chapter is deterministically obtaining positions $\underline{s}_i$ $(i = 1, ..., m)$ for optimally approximating the Gaussian mixture distribution by a deterministic positioning of these samples. The resulting distribution described by

$$f_{DM}(\underline{x}; \mathbf{S}) = \sum_{i=1}^{m} w_i \cdot \delta(\underline{x} - \underline{s}_i) \ ,$$

where $\mathbf{S} = (\underline{s}_1, ..., \underline{s}_M)$.

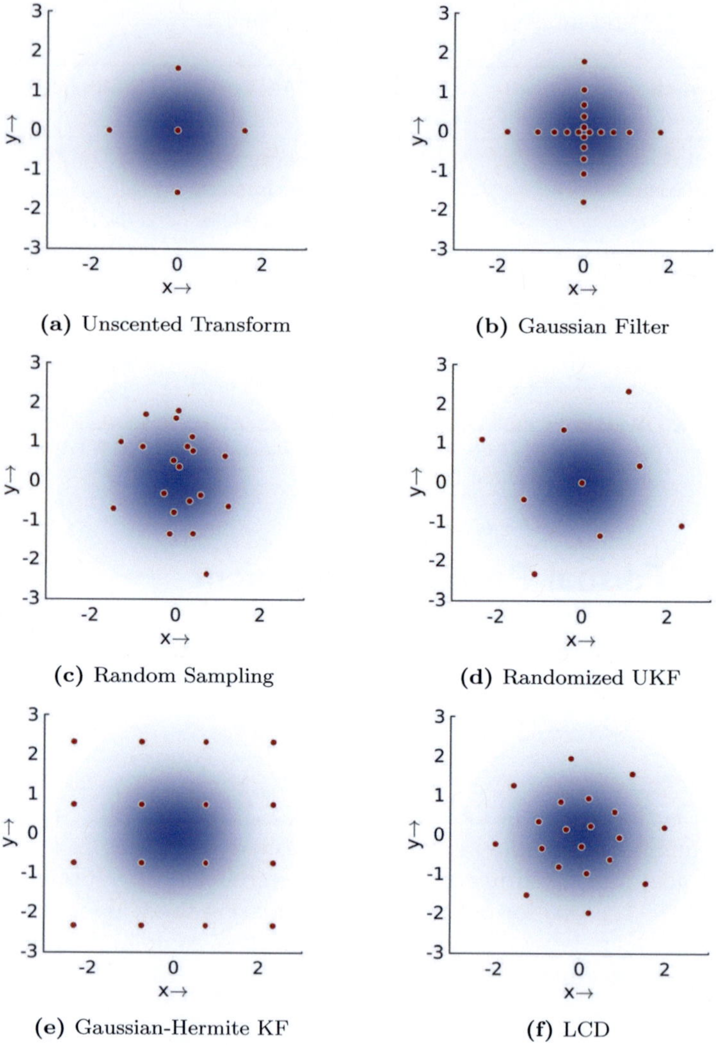

Figure 4.2.: Different sampling schemes.

We are therefore interested in computing

$$\mathbf{S}^* = \arg\min_{\mathbf{S}} d_{w,K}\left(f_{GM}(\underline{x}),\ f_{DM}(\underline{x};\mathbf{S})\right) \ ,$$

where K is a Gaussian RBF kernel and the weighting function is given by

$$w(b) = \begin{cases} b^{-n+1} & b \in [0, b_{max}) \ , \\ 0 & \text{otherwise} \end{cases}$$

In the following, we will discuss this general problem and the important special case of obtaining an approximation of a standard Gaussian. The latter will give rise to an algorithm for generally approximating Gaussian densities.

4.4 Distance Measure for Gaussian Mixtures

For computing the LCD distance measure between a Gaussian mixture and an Dirac mixture, we first need their respective LCDs. For the Dirac mixture the LCD is easily obtained as

$$\mathcal{L}[f_{DM}(\underline{x})](\underline{c}, b) = \sum_{i=1}^{M} K(\underline{c} - \underline{s}_i) \ .$$

The LCD of the Gaussian mixture is a generalization of the LCD of a single axis-aligned Gaussian as derived in [43]. This results in

$$\begin{aligned} \mathcal{L}[f_{GM}](\underline{c}, b) =& \mathbb{E}(K(\underline{c} - \underline{x}, b)) \\ =& \int_{\mathbb{R}^n} K(\underline{c} - \underline{x}, b) \cdot f_{GM}(\underline{x}) \ \mathrm{d}\underline{x} \\ =& \sum_{i=1}^{L} p_i \int_{\mathbb{R}^n} K(\underline{c} - \underline{x}) \cdot f_i(\underline{x}) \ \mathrm{d}\underline{x} \\ =& \sum_{i=1}^{L} p_i \frac{b^n \exp\left(-\frac{1}{2}\left(\underline{c} - \underline{\mu}_i\right)^\top \left(\mathbf{\Sigma}_i + b^2\,\mathbf{I}\right)^{-1} \left(\underline{c} - \underline{\mu}_i\right)\right)}{\sqrt{\det\left(\mathbf{\Sigma}_i + b^2\,\mathbf{I}\right)}} \end{aligned}$$

These representations can now be used directly for computing the LCD distance. However, this requires the computation of multivariate integrals which, when done numerically, is infeasible for most applications. Thus, our first objective is reducing the number of integrals as far as possible.

Proposition 4.2. *The LCD distance $d(f_{GM}, f_{DM})$ can be represented as*

$$d_{w,K}(f_{GM}, f_{DM}) = \int_{\mathbb{R}+} w \cdot (I_1 - 2\, I_2 + I_3)\ \mathrm{d}b \qquad (4.1)$$

where

$$\begin{aligned} I_1 =& \sum_{i=1}^{l} \sum_{j=1}^{l} p_i\, p_j \frac{\left(\sqrt{2\pi}\, b^2\right)^n}{\sqrt{\det(\mathbf{\Sigma}_i + \mathbf{\Sigma}_j + 2b^2 \mathbf{I})}} \\ & \times \exp\left(-\frac{1}{2} (\underline{\mu}_i - \underline{\mu}_j)^\top (\mathbf{\Sigma}_i + \mathbf{\Sigma}_j + 2b^2\mathbf{I})^{-1} (\underline{\mu}_i - \underline{\mu}_j) \right) , \\ I_2 =& \sum_{i=1}^{l} \sum_{j=1}^{m} p_i\, w_j \frac{\left(\sqrt{2\pi}\, b^2\right)^n}{\sqrt{\det(\mathbf{\Sigma}_i + 2b^2 \mathbf{I})}} \\ & \times \exp\left(-\frac{1}{2} (\underline{\mu}_i - \underline{s}_j)^\top (\mathbf{\Sigma}_i + 2b^2\mathbf{I})^{-1} (\underline{\mu}_i - \underline{s}_j) \right) , \\ I_3 =& \sum_{i=1}^{m} \sum_{j=1}^{m} w_i\, w_j\, \pi^{n/2}\, b^n \\ & \times \exp\left(-\frac{1}{2} (\underline{s}_i - \underline{s}_j)^\top (2b^2\mathbf{I})^{-1} (\underline{s}_i - \underline{s}_j) \right) . \end{aligned}$$

Proof. The proof is based in decomposing the original integral into three parts and computing each part individually. This decomposition is obtained by

$$\begin{aligned} d(f_{GM}, p_{DM}) = \int_{\mathbb{R}+} w(b) \Big(& \int_{\mathbb{R}^n} \mathcal{L}[f_{GM}](\underline{c}, b)^2\ \mathrm{d}\underline{c} \\ & - 2 \cdot \int_{\mathbb{R}^n} \mathcal{L}[f_{GM}](\underline{c}, b) \cdot \mathcal{L}[p_{DM}](\underline{c}, b)\ \mathrm{d}\underline{c} \\ & + \int_{\mathbb{R}^n} \mathcal{L}[p_{DM}](\underline{c}, b)^2\ \mathrm{d}\underline{c} \Big)\ \mathrm{d}b \end{aligned}$$

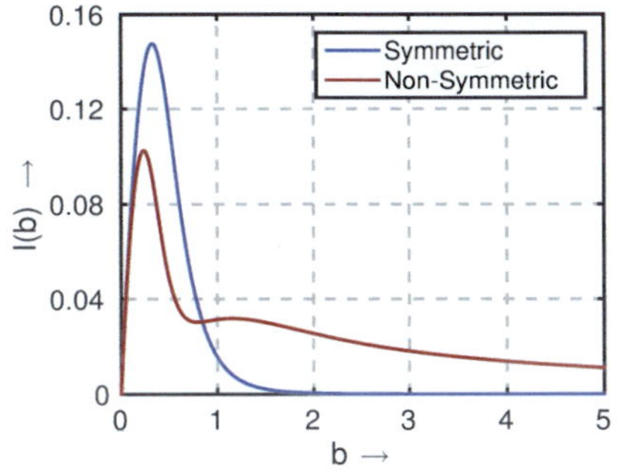

(a) Integrand obtained in Proposition 4.2 for symmetric ($s_1 = -1$, $s_2 = 1$) and non-symmetric ($s_1 = -1$, $s_2 = 0.5$) Dirac positions around the mean.

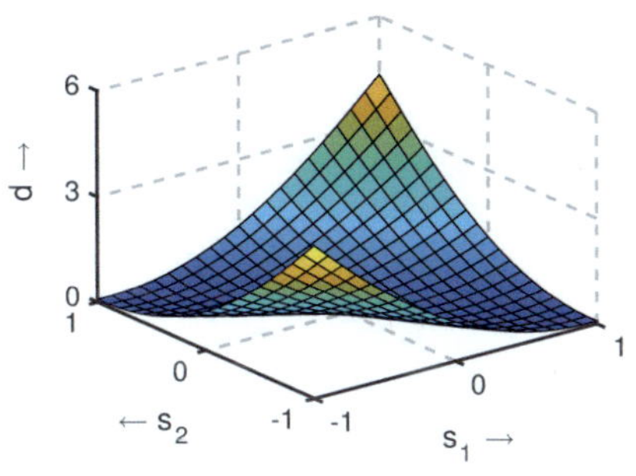

(b) LCD distance for different Dirac positions

Figure 4.3.: Comparing a standard Gaussian with a two-component equally weighted Dirac mixture.

It is now sufficient to show, that the three inner integrals correspond to $I_1(b)$ to $I_3(b)$. For reasons of brevity, this will only be done for $I_1(b)$. For computing $I_1(b)$, we make use of the fact that it is basically an integral over the product of two (not properly normalized) Gaussian densities. Thus, we have

$$\int_{\mathbb{R}^n} \mathcal{L}[f_{GM}](\underline{c}, b)^2 \, \mathrm{d}\underline{c} = \sum_{i=1}^{l} \sum_{j=1}^{l} p_i \, p_j \int_{\mathbb{R}^n} \mathcal{L}[f_i](\underline{c}, b) \cdot \mathcal{L}[f_j](\underline{c}, b) \, \mathrm{d}\underline{c}$$

$$= \sum_{i=1}^{l} \sum_{j=1}^{l} p_i \, p_j \, \left(2 \pi b^2\right)^n \int_{\mathbb{R}^n} \tilde{f}_i(\underline{c}) \cdot \tilde{f}_j(\underline{c}) \, \mathrm{d}\underline{c} \; ,$$

where $\tilde{f}_i(\underline{c})$ is the density of an $\mathcal{N}(\underline{\mu}_i, \mathbf{\Sigma}_i + b^2\mathbf{I})$ distribution. Applying the formula for an integral over the product of two Gaussian densities yields the desired result. Carrying out an analogous procedure for $I_2(b)$ and $I_3(b)$ completes the proof. □

This has removed all but one integral significantly improving the feasibility of actually performing computations involving the considered distance measure. A visualization of this distance and the obtained integrands is shown in Figure 4.3. For many numerical computation procedures the performance can be significantly improved by providing a gradient with respect to the positions of the Dirac delta components. From an algorithmic viewpoint, computation of a gradient is not much different from computing the distance measure itself.

Proposition 4.3. *The derivative of* $d_{w,K}(f_{GM}, f_{DM})$ *with respect to* $s_q^{(r)}$ *is given by*

$$\int_{\mathbb{R}+} w(b)\, G_{q,r}(b)\; \mathrm{d}b = \int_{\mathbb{R}+} w(b)\; (-2\, G_{1,q,r}(b) + G_{2,q,r}(b))\; \mathrm{d}b$$

with

$$\begin{aligned} G_{1,q,r} &= w_q \sum_{i=1}^{l} p_i \frac{(\sqrt{2\pi}\, b^2)^n}{\sqrt{\det(\mathbf{\Sigma}_i + 2b^2\mathbf{I})}} \\ &\quad \times \exp\left(-\frac{1}{2}(\underline{\mu}_i - \underline{s}_q)^\top (\mathbf{\Sigma}_i + 2b^2\mathbf{I})^{-1} (\underline{\mu}_i - \underline{s}_q) \right) , \\ &\quad \times \left((\mathbf{\Sigma}_i + 2b^2\mathbf{I})^{-1} \, (\underline{\mu}_i - \underline{s}_q) \right)^{(r)} , \\ G_{2,q,r} &= w_q \sum_{i=1}^{m} w_i \, \pi^{n/2} \, b^{n-2} \, \exp\left(-\frac{1}{2}(\underline{s}_i - \underline{s}_q)^\top (2b^2\mathbf{I})^{-1} (\underline{s}_i - \underline{s}_q) \right) \\ &\quad \times \left(\underline{s}_i - \underline{s}_q \right)^{(r)} . \end{aligned}$$

Proof. We first observe that due to Lebniz's rule, it is justified to interchange integration and differentiation in $\partial d(f_{GM}, f_{DM})/\partial s_q^{(i)}$ and thus restate this derivative as

$$\int_{\mathbb{R}+} \frac{\partial}{\partial s_q^{(i)}} \left[w(b) \cdot (I_1(b) - 2\, I_2(b) + I_3(b)) \right] \; \mathrm{d}b$$

The term $I_1(b)$ is not considered because it does not involve any $\underline{s}_i$ and thus vanishes after differentiation. Thus, we interpret $I_2(b)$ and $I_3(b)$ as functions of $\underline{s}_i$ and define

$$G_{1,q,r}(b) := \frac{\partial I_2(b)}{\partial s_q^{(r)}} \; , \qquad G_{2,q,r}(b) := \frac{\partial I_3(b)}{\partial s_q^{(r)}} \; .$$

Both derivatives, $G_{1,q,r}$ and $G_{2,q,r}$, are obtained by a straight forward application of the chain rule. For computing $G_{2,q,r}$, we first leave out all terms not involving $\underline{s}_q$. This gives us

$$G_{2,q,r} = 2\, w_q \sum_{i=1}^{m} w_i\, \pi^{n/2}\, b^n \times \frac{\partial}{\partial s_q^{(r)}} \exp\Big(\underbrace{-\frac{1}{2}(\underline{s}_i - \underline{s}_q)^\top (2b^2\mathbf{I})^{-1} (\underline{s}_i - \underline{s}_q)}_{A_{i,q}:=} \Big) .$$

The chain rule gives us

$$\begin{aligned} G_{2,q,r} = 2\, w_q \sum_{i=1}^{m} w_i\, \pi^{n/2}\, b^n & \\ \times \exp\left(-\frac{1}{2}(\underline{s}_i - \underline{s}_q)^\top (2b^2\mathbf{I})^{-1} (\underline{s}_i - \underline{s}_q) \right) & \\ \times \frac{\partial A_{i,q}}{\partial s_q^{(r)}} . & \end{aligned} \tag{4.2}$$

Next, we note that the gradient (with respect to $\underline{x}$) of the product of two vectors $\underline{x}$ and $\underline{c}$ can be obtained directly as

$$\nabla_{\underline{x}} \left(\underline{c}^\top \underline{x} \right) = \nabla_{\underline{x}} \left(\underline{x}^\top \underline{c} \right) = \underline{c} . \tag{4.3}$$

Furthermore, the gradient of a quadratic form when assuming a matrix $\mathbf{A}$ to be symmetric is given by

$$\nabla_{\underline{x}} \left(\underline{x}^\top \, \mathbf{A}\, \underline{x} \right) = 2\, \mathbf{A}\, \underline{x} \tag{4.4}$$

Let $\mathbf{\Sigma}_b := (2b^2\mathbf{I})^{-1}$. Reformulating $A_{i,q}$ as

$$-\frac{1}{2} \left(\underline{s}_q^\top \, \mathbf{\Sigma}_b \, \underline{s}_q - 2 \underline{s}_i^\top \, \mathbf{\Sigma}_b \, \underline{s}_q + \underline{s}_i^\top \, \mathbf{\Sigma}_b \, \underline{s}_i \right)$$

and then using (4.3) and (4.4) we obtain

$$\frac{\partial A_{i,j}}{\partial s_q^{(r)}} = \left(\nabla_{\underline{s}_q} A_{i,q} \right)^{(r)} = \left(\mathbf{\Sigma}_b \left(\underline{s}_i - \underline{s}_q \right) \right)^{(r)} = \frac{1}{2\, b^2} (\underline{s}_i - \underline{s}_q)^{(r)} .$$

Using this, (4.2) can be simplified to the desired result. The corresponding result for $G_{1,q,r}$ is obtained analogously. □

The results of the previous two propositions give rise to an algorithmic implementation of the considered approximation procedure. In the general case of arbitrary Gaussian mixtures, this optimization procedure is nonlinear. Thus, a nonlinear numerical optimizer has to be used for obtaining an approximation. A simplification for the computation of $I_3(b)$ was given in [43] as follows.

Theorem 4.4 (Hanebeck & Huber (2009)). *For large b_{max}, the following closed–form expression for $I_3(b)$ is obtained*

$$\int_{\mathbb{R}+} w \cdot I_3 \text{ db} \approx \frac{\pi^{\frac{n}{2}}}{8} \sum_{i=1}^{m} \sum_{j=1}^{m} w_i w_j \left(4b_{max}^2 - C_b T_{i,j} + xlog(T_{i,j})\right) ,$$

with $xlog(z) = z \cdot \log(z)$ and constants $C_b = \log(4\, b_{max}^2) - \gamma$ (here, γ denotes the Euler-Mascheroni constant) and

$$T_{ij} = \sum_{k=1}^{n} \left(s_i^{(k)} - s_j^{(k)}\right)^2 . \tag{4.5}$$

In the same paper, the methodology to obtain this result was also used for a simplification of the derivative $G_{2,q,r}$.

Theorem 4.5 (Hanebeck & Huber (2009)). *For large b_{max}, a closed–form expression for the second term of the gradient of the distance measure with respect to the locations of the Dirac components is given by*

$$\int_{\mathbb{R}+} w \cdot G_{2,q,r} \text{ d}b \approx \frac{\pi^{n/2}}{2} w_q \left[\sum_{i=1}^{m} w_i \left(\underline{s}_q^{(r)} - \underline{s}_i^{(r)}\right) \log\left(\left|\left|\underline{s}_q - \underline{s}_i\right|\right|_2^2\right) + C_b \left(\sum_{i=1}^{m} w_i\, \underline{s}_i^{(r)} - \underline{s}_q^{(r)}\right)\right] ,$$

where C_b is defined as in the previous theorem.

These two theorems additionally simplify the computation of the distance measure. However, there are still two challenges. First, the entire distance measure involves still a b_{max} parameter, which needs to be chosen properly. Second, there is still an integral, which, so far, has to be solved numerically. In the following, our goal is to investigate special cases where these problems can be avoided.

4.5 Approximation of Even-Dimensional Gaussians

This section addresses the problem of numerical integration in (4.1) for computation of the LCD distance. In the even-dimensional case, the numerical integration can be replaced by well known mathematical functions, particularly the exponential integral, which can be implemented using precomputed lookup tables or based on its series expansion.

Thus, the first two terms from Proposition 4.2 simplify to

$$
\begin{aligned}
I_1 &= \frac{\left(\sqrt{\pi}\, b^2\right)^n}{\sqrt{1+b^2}} \,, \\
I_2 &= \sum_{i=1}^{l} \sum_{j=1}^{m} p_i\, w_j \, \frac{\left(\sqrt{2\pi}\, b^2\right)^n}{\sqrt{\det(\mathbf{\Sigma}_i + 2b^2 \mathbf{I})}} \\
&\qquad \times \exp\left(-\frac{1}{2} (\underline{\mu}_i - \underline{s}_j)^\top (\mathbf{\Sigma}_i + 2b^2\mathbf{I})^{-1} (\underline{\mu}_i - \underline{s}_j) \right) \,.
\end{aligned}
$$

In the proof of Theorem 4.4, the following representation for the integral of $w \cdot I_3$ was obtained

$$
\int_{\mathbb{R}+} w \cdot I_3 \, \mathrm{d}b = \pi^{\frac{n}{2}} \sum_{i=1}^{m} \sum_{j=1}^{m} w_i w_j \, C(b_{max}, T_{i,j})
$$

with

$$
C(b,c) = \frac{b^2}{2} \exp\left(-\frac{1}{2} \frac{c}{2b^2} \right) + \frac{c}{8} \,\mathrm{Ei}\left(-\frac{1}{2} \frac{c}{2b^2} \right) \tag{4.6}
$$

This representation is exact for arbitrary positive b_{max}. However, it involves the exponential integral $\mathrm{Ei}(x)$, which needs to be either precomputed or evaluated using a series expansion. A definition is given in appendix A.1 and an algorithm for its computation can be found in [21]. For the gradient component $G_{2,q,r}$ we have an analogous situation, that is

$$
\int_{\mathbb{R}+} w \cdot G_{2,q,r} \, \mathrm{d}b = \frac{\pi^{\frac{n}{2}}}{2} p_q \sum_{i=1}^{m} p_i (s_q^{(r)} - s_i^{(r)}) \,\mathrm{Ei}\left(-\frac{1}{2} \frac{T_{q,i}}{2b_{max}^2} \right) \,.
$$

It remains to derive similar representations for integrals of $w \cdot I_1$, $w \cdot I_2$, and $w \cdot G_{1,q,r}$. The proofs given below were originally derived in a slightly modified form in [135]. In the first case, a very general result is possible.

Theorem 4.6. *For the case of an n-dimensional standard normal Gaussian, the integral of $w \cdot I_1$ is obtained as*

$$\int_{\mathbb{R}+} w \cdot I_1 \, \mathrm{d}b = \pi^{n/2} A_n(b_{max})$$

where

$$\begin{aligned} A_1(b) =& \frac{1}{2}\left(b\sqrt{1+b^2} - \operatorname{arcsinh}(b)\right) , \\ A_2(b) =& \frac{1}{2}\left(b^2 - \log(1+b^2)\right) , \\ A_{2k}(b) =& 2k\left(\frac{A_2(b)}{2} - \sum_{i=2}^{k-1} \frac{b^{2i}(\sqrt{1+b^2})^{2-2i}}{2i(2i-2)}\right) - \frac{t^{2k}(\sqrt{1+t^2})^{2-2k}}{(2k-2)} , \\ A_{2k-1}(b) =& (2k-1)\left(A_1(b) - \sum_{i=1}^{k-2} \frac{t^{2i+1}(\sqrt{1+b^2})^{1-2i}}{4i^2-1}\right) \\ & - \frac{b^{2k-1}(\sqrt{1+b^2})^{3-2k}}{(2k-3)} , \end{aligned} \tag{4.7}$$

for $k \in \mathbb{N}$ and $k \geq 2$.

Proof. The proof is subdivided three steps. In the first two steps we compute the base cases A_1 and A_2 respectively. Then, we obtain a recursion formula that gives us the result for A_{2k} and A_{2k-1}.

Computation of A_1

The considered integral can be separated using $f(b) = b/2$ and $g(b) = \sqrt{1+b^2}$ as follows

$$A_1(b) = \int_0^b \frac{x^2}{\sqrt{1+2x^2}} \, \mathrm{d}x = \int_0^b f(x)g'(x) \, \mathrm{d}x$$

Applying integration by parts yields

$$A_1(b) = \left[\frac{1}{2}x\sqrt{1+x^2}\right]_0^b - \int_0^b \frac{1}{2\sqrt{1+x^2}}$$
$$= \frac{1}{2}b\sqrt{1+b^2} - \frac{\operatorname{arcsinh}(b)}{2} \ .$$

Computation of A_2

For computing A_2 we make use of the fact that

$$\frac{\partial \ln(1+x^2)}{\partial x} = \frac{x}{1+x^2} \ . \tag{$*$}$$

Now, we compute directly

$$A_2(b) = \int_0^b \frac{x^3}{1+x^2}\,\mathrm{d}x = \int_0^b \frac{x(x^2+1-1)}{1+x^2}\,\mathrm{d}x$$
$$= \int_0^b x - \frac{x}{1+x^2}\,\mathrm{d}x \overset{(*)}{=} \frac{b^2}{2} - \frac{1}{2}\log(1+b^2) \ .$$

Derivation of Recursion formula for $n > 2$

Once again, integration by parts is used. Therefore, we define $f(x) = \sqrt{1+x^2}$ and $g(x) := x^n(1+x^2)^{(-n+1)/2}$. The derivatives of these functions are given by

$$f'(x) = \frac{x}{\sqrt{1+x^2}}$$
$$g'(x) = \frac{n\,x^{n-1}}{(\sqrt{1+x^2})^{n-1}} - \frac{(n-1)x^{n+1}}{(\sqrt{1+x^2})^{n-1}} \ .$$

The integrand can be separated into $f'(x)g(x)$. This gives us

$$A_n(b) = \int_0^b \frac{x^{n+1}}{\left(\sqrt{1+x^2}\right)^n}\,\mathrm{d}x = \int_0^b f'(x)\,g(x)\,\mathrm{d}x$$
$$= \left[\frac{x^n}{(\sqrt{1+x^2})^{n-2}}\right]_0^b - \int_0^b \frac{n\,x^{n-1}}{(\sqrt{1+x^2})^{n-2}} - \frac{(n-1)\,x^{n+1}}{(\sqrt{1+x^2})^n}\,\mathrm{d}x$$
$$= \frac{b^n}{(\sqrt{1+b^2})^{n-2}} - n\,A_{n-2}(b) + (n-1)\,A_n(b) \ .$$

This is used to compute a recursion formula

$$A_n(b) = \frac{1}{n-2}\left(n\,A_{n-2}(b) - \frac{b^n}{(\sqrt{1+b^2})^{n-2}}\right)\ .$$

Formulas for $A_{2k}(b)$ and $A_{2k-1}(b)$ are obtained by resolving this recursion using an induction argument. □

A similar result was also obtained for $w \cdot I_2$. However, this is unfortunately only possible for even-dimensional scenarios.

Theorem 4.7. *In an even-dimensional scenario with $n = 2\,k$ for $k \in \mathbb{N}$, the integral of $w \cdot I_2$ is given by*

$$\int_{\mathbb{R}+} w \cdot I_2 \;\mathrm{d}b = (2\pi)^{n/2} \sum_{i=1}^{m} w_i \left(B_{k,k}\left(b_{max}, ||\underline{s}_i||_2^2\right) - B_{k,k}\left(0, ||\underline{s}_i||_2^2\right)\right)$$

with

$$B_{k,k}(b,c) = \frac{1}{2^k}\sum_{j=0}^{k}(-1)^j\binom{k}{j}B_{0,j}(b,c)\ ,$$

and the base cases

$$\begin{aligned} B_{0,0}(b,c) =& \frac{1+2b^2}{4}\exp\left(-\frac{c}{2+4b^2}\right) + \frac{c}{8}\,\mathrm{Ei}\left(-\frac{c}{2+4b^2}\right)\ , \\ B_{0,1}(b,c) =& -\frac{1}{4}\,\mathrm{Ei}\left(-\frac{c}{2+4b^2}\right)\ , \\ B_{0,d}(b.c) =& \exp\left(-\frac{c}{2+4b^2}\right) \\ & \times \sum_{j=2}^{d}\frac{(d-2)!\;2^{d-j-1}}{(j-2)!\;c^{d-j+1}\;(1+2b^2)^{j-2}}\ , \end{aligned} \tag{4.8}$$

for $d \geq 2$.

Proof. Every summand in $w \cdot I_2$ is of the form

$$w_i\,\frac{(2\pi)^k\,b^{2k+1}}{(1+2b^2)^k}\exp\left(-\frac{1}{2}\frac{c}{1+2b^2}\right)\ ,$$

with some $c > 0$. Thus, it is sufficient to compute the antiderivative of this function. Once again, this will be done in three steps. First, a very general recursive relationship will be derived. Then, a recursive relationship for the base case $B_{0,d}$ when $d \geq 2$. Finally, the base cases $B_{0,0}$ and $B_{0,1}$ will be computed.

Derivation of a general recursion
First, we observe that for $u, v \in \mathbb{N}$ it holds

$$\begin{aligned}\frac{b^{2u+1}}{(1+2b^2)^v} &= \frac{2b^{2u+1}+b^{2u-1}-b^{2u-1}}{2(1+2b^2)^v}\\ &= \frac{b^{2u-1}(1+2b^2)-b^{2u-1}}{2(1+2b^2)^v}\\ &= \frac{b^{2u-1}}{2(1+2b^2)^{v-1}} - \frac{b^{2u-1}}{2(1+2b^2)^v} \ .\end{aligned}$$

This motivates the definition

$$\frac{\partial B_{u,v}(b,c)}{\partial b} := \frac{b^{2k+1}}{(1+2b^2)^k} \exp\left(-\frac{1}{2}\cdot\frac{c}{1+2b^2}\right)$$

and yields the following recursion

$$B_{u,v}(b,c) = \frac{1}{2}\left(B_{u-1,v-1}(b,c) - B_{u-1,v}(b,c)\right) \ . \tag{$*$}$$

This resembles the summation scheme in Pascal's triangle. Thus, a corresponding representation of this recursion yields the result for $B_{k,k}$.

Derivation of base cases $B_{0,d}$ for $d \geq 2$
First, we only consider the case where $d \geq 3$. In order to use integration by parts, we define

$$f_d(x,c) := \frac{1}{2c(1+2x^2)^{d-2}} \ , \quad g(x,c) := \exp\left(\frac{c}{2+4x^2}\right)$$

with partial derivatives (with respect to x)

$$\frac{\partial f_d(x,c)}{\partial x} = \frac{1}{2c(1+2x^2)^{d-2}} \ , \quad \frac{\partial g(x,c)}{\partial x} = \exp\left(\frac{c}{2+4x^2}\right) \ .$$

Thus, we have

$$\begin{aligned} B_{0,d}(b,c) - B_{0,d}(0,c) &= \int_0^b f_d(x)\, \frac{\partial g'(x,c)}{\partial x}\, \mathrm{d}x \\ &= \left[\frac{\exp\left(-\frac{c}{2+4x^2}\right)}{2c(1+2x^2)^{d-2}} \right]_0^b \\ &\quad - \int_0^b \frac{4\,x(2-d)\exp\left(-\frac{c}{2+4\,x^2}\right)}{2c(1+2\,x^2)^{d-1}}\, \mathrm{d}x\ . \end{aligned}$$

This yields the recursive equation

$$B_{0,d}(b,c) = \left(\frac{\exp\left(-\frac{c}{2+4b^2}\right)}{2c(1+2b^2)^{d-2}} \right) + \frac{2(d-2)}{c} B_{0,d-1}(b,c)\ .$$

Computation of the base case $B_{0,2}(b,c)$ is straightforward

$$\begin{aligned} B_{0,2}(b,c) - B_{0,2}(0,c) &= \int_0^b \frac{x}{2+4x^2} \exp\left(-\frac{c}{(1+2x^2)^2}\right)\, \mathrm{d}x \\ &= \frac{1}{2c} \int_0^b \frac{2cx}{(1+2x^2)^2} \exp\left(-\frac{c}{2+4x^2}\right)\, \mathrm{d}x \\ &= \frac{1}{2c} \int_0^b \frac{8cx}{(2+4x^2)^2} \exp\left(-\frac{c}{2+4x^2}\right)\, \mathrm{d}x \\ &= \left[\frac{1}{2c} \exp\left(-\frac{c}{2+4x^2}\right) \right]_0^b\ . \end{aligned}$$

Using this base case, the recursion above can be resolved, which yields our formula for $B_{0,j}(b)$ for $j \geq 2$.

Computation of $B_{0,1}(b,c)$
We use substitution to compute the antiderivative $B_{0,1}(b)$. Therefore, we define $f(b,c) := -c(2+4b^2)^{-1}$. Using this definition, we obtain

$$B_{0,1}(b,c) - B_{0,1}(0,c) = \int_0^b \frac{x}{1+2x^2} \exp\left(-\frac{c}{2+4x^2}\right)\, \mathrm{d}x$$

$$= -\frac{1}{4}\int_0^b f(x,c)\left(\frac{\partial f(y,c)}{\partial y}\right)_{y=x} \exp(\varphi(x))\;\mathrm{d}x$$
$$= -\frac{1}{4}\int_{f(0,c)}^{f(b,c)} \frac{\mathrm{e}^x}{x}\;\mathrm{d}x$$
$$= -\frac{1}{4}\left[\mathrm{Ei}\left(-\frac{c}{2+4x^2}\right)\right]_0^b .$$

Computation of $B_{0,0}(b,c)$
We show that $B_{0,0}$ can be expressed in terms of $B_{0,1}$ and $B_{0,2}$. Integration by parts is used as a first step. Thus, we define $f(b) = b^2/2$ and $g(b,c) = \exp(-c(2+4b^2)^{-1})$ and obtain

$$B_{0,0}(b,c) - B_{0,0}(0,c) = \int_0^b f'(x)g(x,c)\;\mathrm{d}x$$
$$= \left[\frac{x^2}{2}\exp\left(-\frac{c}{2+4x^2}\right)\right]_0^b$$
$$- \int_0^b \frac{8\,c\,x^3}{2(2+4x^2)^2}\exp\left(-\frac{c}{2+4x^2}\right)\;\mathrm{d}x\;.$$

The integrand of the remaining integral is $c \cdot \partial B_{1,2}(b,c)/\partial b$. Using $(*)$, we get

$$B_{0,0}(b,c) = \frac{b^2}{2}\exp\left(-\frac{c}{2+4b^2}\right) - \frac{c}{2}B_{0,1}(b,c) + \frac{c}{2}B_{0,2}(b,c)$$
$$= \frac{1+2b^2}{4}\exp\left(-\frac{c}{2+4b^2}\right) + \frac{c}{8}\,\mathrm{Ei}\left(-\frac{c}{2+4b^2}\right)\;,$$

which completes the proof. □

The integral of $w \cdot G_{1,q,r}$ is computed in a similar way. The proof is omitted for brevity and we only state the following result.

Proposition 4.8. *In an even-dimensional scenario with* $n = 2\,k$ *for* $k \in \mathbb{N}$, *the integral* $w \cdot G_{1,q,r}$ *is given by*

$$\int_{\mathbb{R}^+} w \cdot G_{1,q,r}\;\mathrm{d}b = s_q^{(r)}\left(B_{k,k+1}(b_{max}, ||\underline{s}_q||_2^2) - B_{k,k+1}(0, ||\underline{s}_q||_2^2)\right)$$

with

$$B_{k,k+1}(b,c) = \frac{1}{2^k} \sum_{j=0}^{k} (-1)^j \binom{k}{j} B_{0,j+1}(b,c) \ ,$$

where the base cases $B_{0,i}(b,c)$ are the same as in the previous proof.

This results can be used directly to approximating standard Gaussian densities. Use of efficient algorithms computing $\mathrm{Ei}(\cdot)$ yields faster evaluation of the LCD distance measure for low dimensions. An approximation of an arbitrary non-standard Gaussian distribution $\mathcal{N}(\underline{\mu}, \mathbf{\Sigma})$ is possible using the following relationship. If $\underline{x} \sim \mathcal{N}(\underline{0}, \mathbf{I})$, then $\left(\mathbf{\Sigma}^{1/2}\,\underline{x} + \underline{\mu}\right) \sim \mathcal{N}(\underline{\mu}, \mathbf{\Sigma})$. That is, our approximation of a standard Gaussian can be used to approximate arbitrary Gaussian distributions. Thus, the resulting procedure is given in Algorithm 4.1.

Algorithm 4.1 Approximation of Gaussians

```
procedure DM APPROXIMATION(μ, Σ, m)
    (s*_1, ..., s*_m) ← arg min d_{w,K}(N(0, I), Σ_{i=1}^{m} δ(x − s_i));
    for i ∈ {1, ..., m} do
        s*_i ← Σ^{1/2} s*_i + μ;
    end for
    return (s*_1, ..., s*_m);
end procedure
```

This approximation is suboptimal in the sense that it does not minimize the LCD distance measure, because this measure is in general not invariant under transformations involving multiplications of a random vector with a matrix square-root of a covariance matrix. This has also some impact on the propagation properties of the LCD as will be seen in the evaluations. However, this sub-optimality can be dealt with by choosing an even higher number of samples.

The proposed method still requires the use of some numerical techniques or precomputed lookup tables for the evaluation of $\mathrm{Ei}(\cdot)$ and a minimization procedure to obtain an optimal approximation of a standard Gaussian. However, this approach might be infeasible in real-time applications. In these scenarios, it is still possible to precompute an approximation of standard Gaussians and then apply the proposed algorithm without the numerical optimization step. This methodology can be generalized to approximation of Gaussian mixtures by approximating each component individually. The fact, that the number of samples may be chosen arbitrarily can compensate for the suboptimality of this approach.

4.6 Avoiding Choice of b_{max} in the Case of Equal Means

So far, our method still requires choosing a suitable b_{max}. Thus, our goal is the derivation of a method avoiding this choice. An algebraically tedious look at the series expansions suggests that letting $b_{max} \to \infty$ is possible in the case of equal means of the Gaussian and the Dirac mixture. That is, with our combined choice of kernel and weighting function the integral over b in our distance measure d converges only in this case.

We will also need the series expansion of the exponential which is

$$\exp(x) = \sum_{i=0}^{\infty} \frac{x^i}{i!} \ . \tag{4.9}$$

Finally, the following Lemma will be helpful in the course of the proof.

Lemma 4.9. *For arbitrary $k \in \mathbb{N}$ with $k \geq 2$, the following identity holds*

$$\frac{k}{2} - \frac{1}{2k-2} - 2k \sum_{i=2}^{k-1} \frac{1}{2i(2i-2)} = \frac{1}{2} \ . \tag{4.10}$$

Proof. First, we restate the following simple identity.

$$\frac{k}{2} - \frac{1}{2k-2} = \frac{k^2 - k - 1}{2k-2} \tag{4.11}$$

Second, it holds for all $k \in \mathbb{N}$ with $k \geq 2$

$$\sum_{i=2}^{k-1} \frac{1}{i(i-1)} = \frac{k-2}{k-1} \ , \tag{4.12}$$

which can be seen by an induction argument. Obviously, this is true for $k = 2$. If $k > 2$, then

$$\sum_{i=1}^{k-1} \frac{1}{i(i-1)} = \sum_{i=1}^{k-2} \frac{1}{i(i-1)} + \frac{1}{(k-1)(k-2)} = \frac{k-3}{k-2} + \frac{1}{(k-1)(k-2)} = \frac{k-2}{k-1}$$

and, thus, (4.12) holds. With these identities, the left hand side of (4.10) can be finally formulated as

$$\begin{aligned} \frac{k}{2} - \frac{1}{2k-2} - 2k \sum_{i=2}^{k-1} \frac{1}{2i(2i-2)} &\overset{(4.11)}{=} \frac{k^2-k-1}{2k-2} - 2k \sum_{i=2}^{k-1} \frac{1}{2i(2i-2)} \\ &= \frac{k^2-k-1}{2k-2} - \frac{k}{2} \sum_{i=2}^{k-1} \frac{1}{i(i-1)} \\ &\overset{(4.12)}{=} \frac{k^2-k-1}{2k-2} - \frac{k^2-2k}{2k-2} \\ &= \frac{1}{2} \,. \end{aligned}$$

□

Now, we have everything we need to proof the main result.

Theorem 4.10. *In the case of equal means and even dimensions $n = 2k$, we obtain the LCD distance between a standard Gaussian and a Dirac mixture for $b_{max} \to \infty$ as*

$$\begin{aligned} \pi^{-k} D = \lim_{b \to \infty} \Bigg[& k \sum_{i=2}^{k-1} \frac{1}{2i} \\ & - 2 \sum_{i=1}^{m} w_i \left(\frac{c_i + 2k}{8} \ln\left(\frac{c_i}{4}\right) + \frac{k\gamma}{4} - B_{k,k}(0, c_i) \right. \\ & \qquad \left. + \sum_{j=2}^{k} (-1)^j \binom{k}{j} \frac{(j-2)!\, 2^{j-3}}{c_i^{j-1}} \right) \\ & + \sum_{i,j} w_i\, w_j \frac{T_{i,j}}{8} \ln\left|\frac{T_{i,j}}{4}\right| \Bigg] \,. \end{aligned} \tag{4.13}$$

with

$$c_i = ||\underline{s}_i||_2^2 \; , \quad T_{i,j} = \sum_{k=1}^{n} (\underline{s}_i^{(k)} - \underline{s}_j^{(k)})^2 \; .$$

Proof. The proof is carried out by investigating which terms in the integrals over $w \cdot I_1$, $w \cdot I_2$, and $w \cdot I_3$ converge as b goes to ∞ (they will be usually denoted by r) and then making use of the fact that the mean of the Dirac mixture is 0.

Consideration of $w \cdot I_1$
For the integral of $w \cdot I_1$, we first consider the case of $n = 2$ using (4.7) and observe that

$$\begin{aligned}
0 &= \left(A_2(b) - \frac{b^2 - \ln(1+b^2)}{2} \right) \\
&= \left(A_2(b) - \frac{b^2 - \ln(b^{-2}+1) - \ln(b^2)}{2} \right) \\
&= \lim_{b \to \infty} \underbrace{\left(A_2(b) - \frac{b^2 - \ln(b^2)}{2} \right)}_{r_{1,1}(b):=} ,
\end{aligned}$$

Using (4.7) again, we can rewrite the integral of $w \cdot I_1$ for even dimension $n \geq 4$ (and thus $k \geq 2$) as

$$\begin{aligned}
\pi^{-n/2} \int w(b) \cdot I_1(b) \, \mathrm{d}b =& A_{2k}(b) \\
=& 2k \left(\frac{A_2(b)}{2} - \sum_{i=2}^{k-1} \frac{b^{2i}(\sqrt{1+b^2})^{2-2i}}{2i(2i-2)} \right) \\
& - \frac{b^{2k}(\sqrt{1+b^2})^{2-2k}}{(2k-2)} \\
=& 2k \left(\frac{1}{4} \left(b^2 - \ln(1+b^2) \right) - \sum_{i=2}^{k-1} \frac{b^{2i}(1+b^2)^{1-i}}{2i(2i-2)} \right) \\
& - \frac{b^{2k}(1+b^2)^{1-k}}{(2k-2)}
\end{aligned}$$

For a later term by term investigation, we will make use of the following identities. First, by letting $i >= 2$, we obtain

$$\begin{aligned}\frac{b^{2i}(1+b^2)^{1-i}}{2i(2i-2)} - \frac{b^2}{2i(2i-2)} =& \frac{b^{2i}(1+b^2) - b^2(1+b^2)^i}{2i(2i-2)(1+b^2)^i} \\ =& \frac{b^{2i} + b^{2i+2} - \left[b^2 \sum_{j=0}^{i} \binom{i}{j} b^{2j}\right]}{2i(2i-2)(1+b^2)^i} \\ =& \frac{b^{2i}(1-i)}{2i(2i-2)(1+b^2)^i} + \mathcal{O}\left(\frac{1}{b^2}\right) \\ =& -\frac{b^{2i}}{4i(1+b^2)^i} + \mathcal{O}\left(\frac{1}{b^2}\right),\end{aligned} \tag{4.14}$$

which converges to $-(4i)^{-1}$ as $b \to \infty$. The second consideration is similar

$$\begin{aligned}\frac{b^{2k}(1+b^2)^{1-k}}{(2k-2)} - \frac{b^2}{2k-2} =& \frac{b^{2k}(1+b^2) - b^2(1+b^2)^k}{(2k-2)(1+b^2)^k} \\ =& \frac{b^{2k}(1-k)}{(2k-2)(1+b^2)^k} + \mathcal{O}\left(\frac{1}{b^2}\right) \\ =& -\frac{b^{2k}}{2(1+b^2)^k} + \mathcal{O}\left(\frac{1}{b^2}\right)\end{aligned} \tag{4.15}$$

which converges to -0.5 as $b \to \infty$. Now, we are ready to consider $A_{2k}(b)$. We make use of our result for $A_2(b)$ and rewrite the terms as

$$\begin{aligned}0 =& \left(A_{2k}(b) - 2k\left(\frac{1}{4}\left(b^2 - \ln(1+b^2)\right) - \sum_{i=2}^{k-1} \frac{b^{2i}(1+b^2)^{1-i}}{2i(2i-2)}\right)\right. \\ &\left. + \frac{b^{2k}(1+b^2)^{1-k}}{(2k-2)}\right) \\ =& \lim_{b\to\infty}\left(A_{2k}(b) - \frac{k}{2}\left(b^2 - \ln(b^2)\right) + 2k\sum_{i=2}^{k-1} \frac{b^{2i}(1+b^2)^{1-i}}{2i(2i-2)}\right. \\ &\left. + \frac{b^{2k}(1+b^2)^{1-k}}{(2k-2)}\right)\end{aligned}$$

Now, we use Lemma 4.9 in order to obtain

$$\begin{aligned}
0 = \lim_{b\to\infty} \Bigg(& A_{2k}(b) - \frac{k}{2}\left(b^2 - \ln(b^2)\right) + 2k\sum_{i=2}^{k-1} \frac{b^{2i}(1+b^2)^{1-i}}{2i(2i-2)} \\
& + \frac{b^{2k}(1+b^2)^{1-k}}{(2k-2)} \Bigg) \\
= \lim_{b\to\infty} \Bigg(& A_{2k}(b) + \frac{b^2}{2} - \frac{k}{2}\ln(b^2) \\
& + 2k\sum_{i=2}^{k-1} \left(\frac{b^{2i}(1+b^2)^{1-i}}{2i(2i-2)} - \frac{b^2}{2i(2i-2)} \right) \\
& + \left(\frac{b^{2k}(1+b^2)^{1-k}}{(2k-2)} - \frac{b^2}{2k-2} \right) \\
& - b^2 \underbrace{\left(\frac{k}{2} - \frac{1}{2k-2} - 2k\sum_{i=2}^{k-1} \frac{1}{2i(2i-2)} - \frac{1}{2} \right)}_{=0} \Bigg)
\end{aligned}$$

Due to the convergence (as $b \to \infty$) of (4.14) and (4.15) this simplifies to

$$0 = \lim_{b\to\infty} \underbrace{\left(A_{2k}(b) - \frac{1}{2}\left(b^2 - k\ln(b^2)\right) - k\sum_{i=2}^{k-1} \frac{1}{2i} - \frac{1}{2} \right)}_{r_{1,k}(b):=} .$$

This formula for $r_{1,k}$ only holds for $k > 1$, because there is no summand 0.5 in $r_{1,1}$. However, it can be formulated in its full generality as follows

$$r_{1,k}(b) = A_{2k}(b) - \frac{1}{2}\left(b^2 - k\ln(b^2)\right) - k\sum_{i=2}^{k} \frac{1}{2i} .$$

And we can rewrite $A_{2k}(b)$ as

$$A_{2k}(b) = \frac{b^2}{2} - \frac{k}{2}\ln(b^2) + k\sum_{i=2}^{k} \frac{1}{2i} + r_{1,k}(b) . \tag{4.16}$$

Consideration of $w \cdot I_2$

For the integral of $w \cdot I_2$, we first investigate the asymptotic behavior of $B_{0,d}(b, c)$ (defined as in (4.8)) for $d \geq 2$ as $b \to \infty$. This converges according to

$$0 = \lim_{b\to\infty} \underbrace{\left(B_{0,d}(b,c) - \frac{(d-2)!\, 2^{d-3}}{c^{d-1}} \right)}_{r_{2,k,d}(b,c):=}$$

and thus we have

$$B_{0,d}(b,c) = \frac{(d-2)!\, 2^{d-3}}{c^{d-1}} + r_{2,k,d}(b,c) \; .$$

Next, we consider $B_{0,1}(b, c)$. Using (A.2) we obtain

$$\begin{aligned}
0 &= \lim_{b\to\infty} \left(B_{0,1}(b,c) + \frac{1}{4} \ln\left(\frac{c}{2+4b^2}\right) + \frac{\gamma}{4} \right) \\
&= \lim_{b\to\infty} \left(B_{0,1}(b,c) + \frac{1}{4} \left(\ln\left(\frac{c}{2b^{-2}+4}\right) - \ln(b^2) \right) + \frac{\gamma}{4} \right) \\
&= \lim_{b\to\infty} \underbrace{\left(B_{0,1}(b,c) + \frac{1}{4} \left(\ln\left(\frac{c}{4}\right) - \ln(b^2) \right) + \frac{\gamma}{4} \right)}_{r_{2,k,1}(b,c):=}
\end{aligned}$$

where γ is the Euler-Mascheroni constant ($\gamma \approx 0.5772$). This yields

$$B_{0,1}(b,c) = -\frac{1}{4} \left(\ln\left(\frac{c}{4}\right) - \ln(b^2) + \gamma \right) + r_{2,k,1}(b,c) \; .$$

For $B_{0,0}(b, c)$, we use a similar approach additionally making use of (4.9) and obtain

$$\begin{aligned}
0 &= \lim_{b\to\infty} \left(B_{0,0}(b,c) - \frac{2+4b^2-c}{8} - \frac{c}{8} \left(\ln\left(\frac{c}{2+4b^2}\right) + \gamma \right) \right) \\
&= \lim_{b\to\infty} \underbrace{\left(B_{0,0}(b,c) - \frac{1}{4} - \frac{b^2}{2} - \frac{c}{8} \left(\ln\left(\frac{c}{4}\right) - \ln(b^2) - 1 + \gamma \right) \right)}_{r_{2,k,0}(b,c):=} ,
\end{aligned}$$

which yields

$$B_{0,0}(b,c) = \frac{1}{4} + \frac{b^2}{2} + \frac{c}{8}\left(\ln\left(\frac{c}{4}\right) - \ln(b^2) - 1 + \gamma\right) + r_{2,k,0}(b,c)$$

We use these three observations to define the function

$$r_{2,k}(b,c) := \frac{1}{2^k}\sum_{j=0}^{k}(-1)^j\binom{k}{j} r_{2,k,j}(b,c) \; ,$$

which converges to 0 as $b \to \infty$. This is used to rewrite $B_{k,k}(b,c)$ as

$$\begin{aligned} B_{k,k}(b,c) &= \frac{1}{2^k}\Bigg(\frac{1}{4} + \frac{b^2}{2} + \frac{c}{8}\left(\ln\left(\frac{c}{4}\right) - \ln(b^2) - 1 + \gamma\right) \\ &\quad + \frac{k}{4}\left(\ln\left(\frac{c}{4}\right) - \ln(b^2) + \gamma\right) \\ &\quad + \sum_{j=2}^{k}(-1)^j\binom{k}{j}\frac{(d-2)!\,2^{d-3}}{c^{d-1}}\Bigg) + r_{2,k}(b,c) \qquad (4.17) \\ &= \frac{1}{2^k}\Bigg(\frac{1}{4} + \frac{b^2}{2} + \frac{c+2k}{8}\left(\ln\left(\frac{c}{4}\right) - \ln(b^2) + \gamma\right) - \frac{c}{8} \\ &\quad + \sum_{j=2}^{k}(-1)^j\binom{k}{j}\frac{(j-2)!\,2^{j-3}}{c^{j-1}}\Bigg) + r_{2,k}(b,c) \; . \end{aligned}$$

Consideration of $w \cdot I_3$
For the integral of $w \cdot I_3$, we investigate the asymptotic behavior of $C(b,c)$ as defined in (4.6) for $b \to \infty$. For the summand involving the exponential function, we make use of (4.9) and obtain

$$\lim_{b\to\infty}\left(\frac{b^2}{2}\exp\left(-\frac{1}{2}\frac{c}{2b^2}\right) - \left(\frac{b^2}{2} - \frac{c}{8}\right)\right) = 0 \; .$$

For treating the summand involving the exponential integral, we make use of (A.2) and obtain

$$\lim_{b\to\infty}\left(\frac{c}{8}\,\mathrm{Ei}\left(-\frac{1}{2}\frac{c}{2b^2}\right) - \frac{c}{8}\left(\gamma + \ln\left|\frac{c}{4b^2}\right|\right)\right) = 0 \; .$$

We define

$$\begin{aligned} r_{3,k}(b,c) &:= C(b,c) - \left(\frac{b^2}{2} - \frac{c}{8}\right) - \frac{c}{8}\left(\gamma + \ln\left|\frac{c}{4b^2}\right|\right) \\ &= C(b,c) - \frac{b^2}{2} - \frac{c}{8}\left(\gamma - 1 + \ln\left|\frac{c}{4}\right| - \ln(b^2)\right) . \end{aligned}$$

A consequence of the preceding discussion is the fact that $r_{3,k}(b,c)$ converges to 0 as $b \to \infty$. This definition is now used to rewrite $C(b,c)$ as

$$C(b,c) = \frac{b^2}{2} + \frac{c}{8}\left(\gamma - 1 + \ln\left|\frac{c}{4}\right| - \ln\left(b^2\right)\right) + r_{3,k}(b,c) . \tag{4.18}$$

Computing the Distance Measure
For computation of the entire distance measure, we make use of the fact that both means are equal, i.e., our sample set has mean 0. Using the definition of $T_{i,j}$ and c_i as above, we obtain

$$\begin{aligned} \sum_{i,j} w_i w_j \, T_{i,j} &= \sum_{i,j} w_i w_j \, ||\underline{s}_i - \underline{s}_j||_2^2 = \sum_{i,j} w_i w_j \sum_{d=0}^{n} \left(\underline{s}_i^{(d)} - \underline{s}_j^{(d)}\right)^2 \\ &= \sum_{i,j} w_i w_j \sum_{d=0}^{n} \underbrace{\left(\underline{s}_i^{(d)}\right)^2}_{c_i} - 2\underline{s}_i^{(d)} \underline{s}_j^{(d)} + \underbrace{\left(\underline{s}_j^{(d)}\right)^2}_{c_j} \\ &= \sum_i w_i \, c_i + \sum_j w_j \, c_j \\ &\quad - \sum_{d=0}^{n} 2 \underbrace{\left(\sum_i w_i \, \underline{s}_i^{(d)}\right)}_{0} \cdot \underbrace{\left(\sum_j w_j \, \underline{s}_j^{(d)}\right)}_{0} \\ &= 2 \sum_i w_i \, c_i . \end{aligned} \tag{4.19}$$

Now, (4.16), (4.17), and (4.18) can be brought together, which finally results in

$$D = \int_0^{\infty} w(b) \left(I_1(b) - 2I_2(b) + I_3(b)\right) \, \mathrm{d}b$$

$$= \pi^k \lim_{b\to\infty} \Big(A_{2k}(b) - 2^{k+1} \sum_{i=1}^{m} w_i (B_{k,k}(b, c_i) - B_{k,k}(0, c_i))$$
$$+ \sum_{i,j} w_i w_j \, C(b, T_{i,j})\Big) \, .$$

Leaving out $r_{i,k}$ as they converge to zero anyway yields

$$\pi^{-k} D = \lim_{b\to\infty} \Bigg[\left(\frac{b^2}{2} - \frac{k}{2} \ln(b^2) + k \sum_{i=2}^{k} \frac{1}{2i} \right)$$
$$- 2 \sum_{i=1}^{m} w_i \left(\frac{1}{4} + \frac{b^2}{2} + \frac{c_i + 2k}{8} \left(\ln\left(\frac{c_i}{4}\right) - \ln(b^2) + \gamma \right) \right.$$
$$\left. - \frac{c_i}{8} - B_{k,k}(0, c_i) + \sum_{j=2}^{k} (-1)^j \binom{k}{j} \frac{(j-2)!\, 2^{j-3}}{c_i^{j-1}} \right)$$
$$+ \sum_{i,j} w_i \, w_j \left(\frac{b^2}{2} + \frac{T_{i,j}}{8} \left(\gamma - 1 + \ln\left|\frac{T_{i,j}}{4}\right| - \ln\left(b^2\right) \right) \right) \Bigg] \, .$$

Some of the terms cancel out immediately

$$\pi^{-k} D = \lim_{b\to\infty} \Bigg[k \sum_{i=2}^{k} \frac{1}{2i}$$
$$- 2 \sum_{i=1}^{m} w_i \left(\frac{1}{4} + \frac{c_i}{8} \left(\ln\left(\frac{c_i}{4}\right) - \ln(b^2) + \gamma \right) \right.$$
$$+ \frac{k}{4} \left(\ln\left(\frac{c_i}{4}\right) + \gamma \right) - \frac{c_i}{8} - B_{k,k}(0, c_i)$$
$$\left. + \sum_{j=2}^{k} (-1)^j \binom{k}{j} \frac{(j-2)!\, 2^{j-3}}{c_i^{j-1}} \right)$$
$$+ \sum_{i,j} w_i \, w_j \left(\frac{T_{i,j}}{8} \left(\gamma - 1 + \ln\left|\frac{T_{i,j}}{4}\right| - \ln\left(b^2\right) \right) \right) \Bigg] \, .$$

Now, (4.19) is applied.

$$\pi^{-k}\, D = \lim_{b\to\infty} \Bigg[k\sum_{i=2}^{k} \frac{1}{2i} - 2\sum_{i=1}^{m} w_i \Bigg(\frac{1}{4} + \frac{c_i}{8}\left(\ln\left(\frac{c_i}{4}\right) - \ln(b^2) + \gamma\right) + \frac{k}{4}\left(\ln\left(\frac{c_i}{4}\right) + \gamma\right) - B_{k,k}(0, c_i) - \frac{c_i}{8} + \sum_{j=2}^{k} (-1)^j \binom{k}{j} \frac{(j-2)!\, 2^{j-3}}{c_i^{j-1}} \Bigg) + 2\sum_{i=1}^{m} w_i \left(\frac{c_i}{8}\left(\gamma - 1 - \ln\left(b^2\right)\right)\right) + \sum_{i,j} w_i\, w_j \frac{T_{i,j}}{8} \ln\left|\frac{T_{i,j}}{4}\right| \Bigg] \,.$$

Once again several terms cancel out, which yields the desired result

$$\pi^{-k}\, D = \Bigg[k\sum_{i=2}^{k} \frac{1}{2i} - 2\sum_{i=1}^{m} w_i \Bigg(\frac{1}{4} + \frac{c_i + 2k}{8} \ln\left(\frac{c_i}{4}\right) + \frac{k\gamma}{4} - B_{k,k}(0, c_i) + \sum_{j=2}^{k} (-1)^j \binom{k}{j} \frac{(j-2)!\, 2^{j-3}}{c_i^{j-1}} \Bigg) + \sum_{i,j} w_i\, w_j \frac{T_{i,j}}{8} \ln\left|\frac{T_{i,j}}{4}\right| \Bigg] \,.$$

□

The result might seem somewhat surprising because choosing a sample set without zero mean and computing the distance measure according to (4.13) still yields a finite result. However, this result would not be the true distance measure, because it does not include the terms which disappeared when using the fact that the sample mean is zero. Thus, this result can not be applied to computing the distance measure for cases without zero mean of the deterministic sample set.

4.7 Evaluation and Discussion

In order to evaluate the proposed approximation, we first give an example of the differences between offline sampling and online sampling. This is simply done by approximating a Gaussian mixture with both approaches. Then, the quality of the sampling scheme is investigated based on an approximate propagation example.

4.7.1 Examples for Online and Offline Sampling

We have performed an approximation of two dimensional Gaussian mixtures in three scenarios. These scenarios were chosen in order to visualize the impact of the overlap on the actual approximation. In all scenarios, a three component Gaussian mixture was used. The covariance matrices of the respective components are given by

$$\mathbf{\Sigma}_1 = \begin{pmatrix} 3 & 0 \\ 0 & 0.1 \end{pmatrix} , \quad \mathbf{\Sigma}_2 = \mathbf{R}(\theta_1)\,\mathbf{\Sigma}_1\,\mathbf{R}(\theta_1)^\top , \quad \mathbf{\Sigma}_2 = \mathbf{R}(\theta_2)\,\mathbf{\Sigma}_1\,\mathbf{R}(\theta_2)^\top ,$$

where $R(\theta)$ is a rotation matrix (with angle θ) and $\theta_1 = 60°$, $\theta_2 = 120°$. These covariance matrices were the same for all three scenarios. However, different means were chosen in order to visualize how overlapping affects the approximation. For the first case, all means were chosen $\mu_1 = \mu_2 = \mu_3 = (0,0)^\top$. In the second case the means were chosen as $\mu_1 = (0,1)^\top$, $\mu_2 = \mathbf{R}(60°) \cdot (0,1)^\top$, and $\mu_3 = \mathbf{R}(-60°) \cdot (0,1)^\top$. Finally, in the third scenario, the means were chosen as $\mu_1 = (0,2)^\top$, $\mu_2 = \mathbf{R}(60°) \cdot (0,2)^\top$, and $\mu_3 = \mathbf{R}(-60°) \cdot (0,2)^\top$.

The approximation results are visualized in Figure 4.4. The online variant was generated with 51 equally weighted samples using the `fminunc` method in Matlab, which uses a trust region algorithm. For the offline approach, we approximated each Gaussian mixture component individually. This was done by computing an optimal approximation (with 17 samples) of a standard Gaussian. Multiplying the matrix square root (obtained by using the Cholesky decomposition) of the covariance matrix and a subsequent repositioning resulted in the desired approximation of each component. At first glance, the offline approximation might seem unsatisfying, because its coverage of the state-space looks less homogeneous. However, it still results in a good approximation of the underlying density as will be seen in the following evaluation.

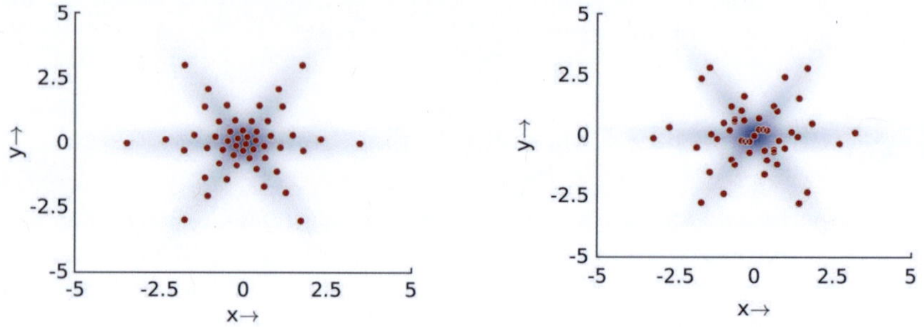

(a) Means $\mu_1 = \mu_2 = \mu_3 = (0,0)^\top$.

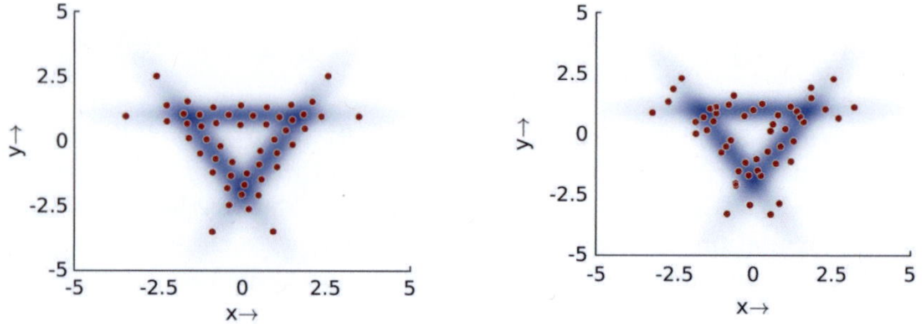

(b) Means $\mu_1 = (0,1)^\top$, $\mu_2 = \mathbf{R}(60°) \cdot (0,1)^\top$, and $\mu_3 = \mathbf{R}(-60°) \cdot (0,1)^\top$.

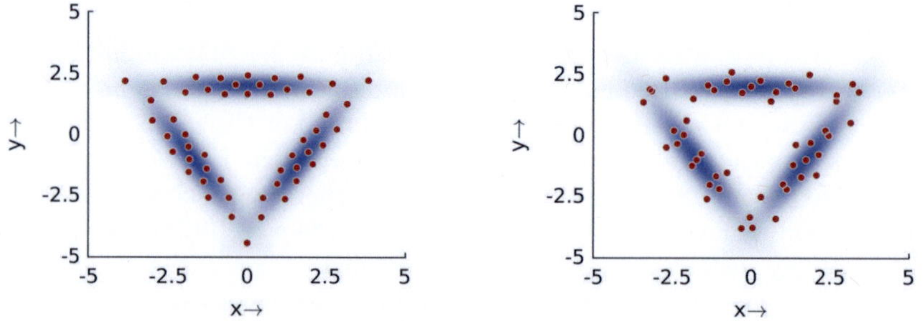

(c) Means $\mu_1 = (0,2)^\top$, $\mu_2 = \mathbf{R}(60°) \cdot (0,2)^\top$, and $\mu_3 = \mathbf{R}(-60°) \cdot (0,2)^\top$.

Figure 4.4.: Comparing the online approach (left column) with the offline approach (right column).

4.7.2 Evaluation of Uncertainty Propagation

In the following, we investigate the quality of the LCD approximation by applying it to approximate propagation. That is, we evaluate the quality of computing $\mathbb{E}(g(\underline{x}))$. As comparison methods, we chose random sampling and the randomized UKF. This choice is due to the fact that these methods have a similarly adaptable number of samples and thus do not suffer from the curse of dimensionality in the same way as the Gauss-Hermite Kalman filter or other grid based numerical integration techniques.

For evaluating the propagation of uncertainty, we considered the function $g(\underline{x}) = \text{atan2}(\underline{x}^{(2)}, \underline{x}^{(1)})7$, where atan2 is the quadrant aware inverse tangent given by

$$\text{atan2}(y, x) = \begin{cases} \arctan(y \cdot x^{-1}) & x > 0 \,, \\ \arctan(y \cdot x^{-1} + \pi) & y \geq 0 \,, x < 0 \,, \\ \arctan(y \cdot x^{-1} - \pi) & y < 0 \,, x < 0 \,, \\ \pi/2 & y > 0 \,, x = 0 \,, \\ -\pi/2 & y < 0 \,, x = 0 \,, \\ \text{undefined} & x = y = 0 \,. \end{cases}$$

This function returns the angle between the positive real axis and the line that connects $\underline{0}$ with $(x, y)^\top$. Thus, it is typically used in bearings-only tracking applications (e.g. in [14]). Furthermore, $\underline{x}$ is assumed to be a Gaussian $\mathcal{N}(\underline{\mu}, \mathbf{C}_i)$ random vector with $\underline{\mu} = (1, 1)^\top$. Two noise scenarios are considered. In the first scenario, the covariance is given by the identity matrix $\mathbf{C}_1 = \mathbf{I}$. In the second scenario the covariance is given by $\mathbf{C}_2 = \text{diag}(3, 0.5)$. The reason for using two scenarios is the fact that we want to evaluate both, the online and the offline approach.

The evaluation is carried out as follows. The LCD based approximation of a standard Gaussian is obtained by minimizing the distance measure. After a repositioning to the desired mean $\underline{\mu}$, the resulting sample set can be directly used in the first scenario. For the second scenario, each sample is multiplied with a matrix square-root of $\mathbf{C}_2$ before repositioning. Thus, the first scenario can be thought of as an online approximation approach whereas the second scenario represents the offline approximation.

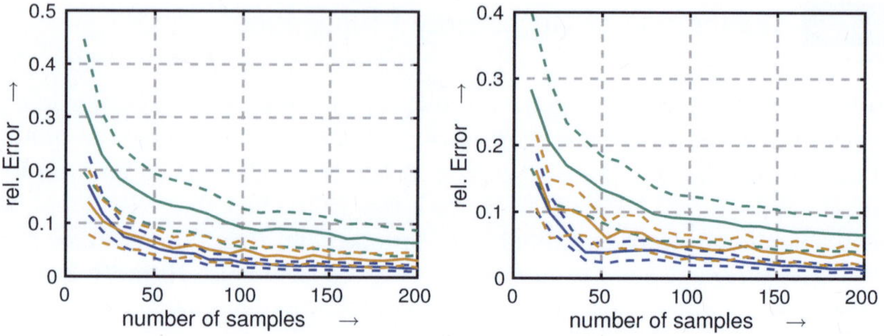

(a) Scenario 1 (Online): $\mathbf{C} = \mathbf{I}$. **(b)** Scenario 2 (Offline): $\mathbf{C} = \text{diag}(3, 0.5)$.

Figure 4.5.: Relative error of propagation for an online (Scenario 1) and an offline (Scenario 2) approximation scenario using a random sampling based approach (green), the randomized UKF (orange), and the LCD based approximation approach (blue). Dashed lines indicate the 0.5σ bounds.

The result of the optimization procedure that we use for obtaining the estimate of a standard Gaussian is dependent on the initial values provided to the optimization procedure. Furthermore, the optimum is invariant under rotation. For this reason, we use different randomly sampled initial values for the optimizer in each run. These initial values are obtained by random sampling from a standard Gaussian. The randomized UKF involves two random sampling steps within each iteration. First, an orthogonal matrix is sampled for randomly rotating an UKF sample set. Second, the UKF sampling set is rescaled randomly.

The results were generated using 100 simulation runs. Ground-truth was obtained using the `integral2` function in Matlab. The relative error is visualized in Figure 4.5. As can be seen, all three approaches converge. However, the LCD based approach has the fastest speed of convergence while simultaneously having the smallest variance. Its main advantage is not limited to this fast convergence speed. More important is the fact that the LCD samples are precomputed in advance which results in a lower computational cost for practical applications.

4.8 Summary and Discussion

This chapter addressed the problem of approximating Gaussians (and their mixtures) by a discrete distribution defined on the same underlying domain. This approximation is of particular interest for stochastic filtering because numerical handling of continuous densities is computationally burdensome. Particularly, linear regression Kalman filters benefit from the proposed approach. The entire approximation procedure is based on obtaining a discrete sample set by minimizing the distance measure from the previous chapter. This is done by a numerical optimization procedure, and thus, requires multiple evaluations of the proposed distance measure.

First, we derived the distance measure for the general case of Gaussian mixtures. Its evaluation is not possible in closed-form and, up to now, it required the choice of a maximum kernel size. To simplify the computation, this chapter proposed a better representation of the distance measure when comparing one single Gaussian with a discrete sample set (which also can be generalized to other cases). Furthermore, it was shown that the choice of a maximum kernel size can be avoided in the case of equal means, which motivates an optimization procedure considering symmetry. Applications in scenarios with real-time requirements and thus not allowing for numerical optimization can be considered by precomputing an approximation of the standard Gaussian and then multiplying with a suitable matrix square-root in order to achieve the desired variance. The evaluations show both approaches to outperform similar state-of-the art approximation techniques.

For future work, it is important to derive error bounds based on the LCD distance measure for the approximate integration procedure. This gives rise to a conceptual advantage of the approach proposed here compared to randomized approaches, which is the fact that the value of the distance measure is known in advance, and thus, the error is not dependent on the outcome of a randomized procedure. The consequence for the use of linear regression Kalman filters based on the LCD approximation would be the ability to obtain deterministic performance guarantees of approximate integration in advance by choosing a sufficient number of samples.

CHAPTER 5

Unscented Orientation Estimation

Contents

5.1 Introduction

So far, our consideration of nonlinear systems focused on considering nonlinearities of the system dynamics. As most real world systems exhibit nonlinear dynamics, the considered scenario was of particular importance. In this chapter, we consider an even more complicated situation by not only assuming nonlinear system dynamics but also nonlinear underlying state spaces.

This is motivated by a big number of applications because many state estimation problems, particularly in robotics or mixed and augmented reality, usually estimate values that are defined on a nonlinear domain. In that instance, estimation in periodic state spaces is of particular importance. Examples of this type of estimation problems involve estimating positions on a ball, states of angular joints, or orientations of objects. In a prepared environment, costly high precision sensors can be used for certain specialized applications. However, the widespread deployment of poor sensors and novel robotic perception applications that involve high uncertainty motivates the development of novel algorithms to deal with these scenarios.

Whenever low levels of uncertainty are involved, estimation algorithms can be derived from the linear case by making use of local linearity. That is, a differentiable manifold can be locally approximated by a linear space. The practical consequence of this fact is a justification for using different types of Kalman Filters for estimating quantities such as angles or orientations. Handling of periodic boundaries is usually done by ad-hoc fixes, such as an intelligent repositioning of measurements. Alternatively, nonlinear projection [56] is used in order to ensure that domain restrictions are satisfied.

However, once high uncertainties are involved, none of these approaches seems promising because they fail to take periodicity into account in their distribution assumption.This also explains why assuming a Gaussian distribution at any point is inherently wrong. One of its main motivations, the central limit theorem, does not hold on (and is not formulated for) periodic domains such as the circle or the hypersphere. In case of the circle, the true limit distribution might be a wrapped normal, which arises by wrapping the one-dimensional Gaussian probability density around an interval of length 2π, or equivalently by taking a Gaussian random variable modulo 2π. Thus, taking the summation scheme from the central limit theorem modulo 2π yields convergence (in distributional sense) to a wrapped normal distribution.

Therefore, we use results from directional statistics [89, 49] for a correct consideration of the underlying domain. This chapter contributes to periodicity-aware estimation algorithms by proposing a novel deterministic sampling scheme for hyperspherical antipodally symmetric distributions based on. The proposed algorithm is based on moment matching and can be thought of as a hyperspherical equivalent of the UKF. This sampling scheme is used to develop a novel filter based on the Bingham distribution, which is suitable for estimation on the hypersphere. In particular, we focus on orientation estimation, where orientation is represented by unit quaternions. One of the challenges involved in using the Bingham distribution is the computation of its normalization constant. We address this challenge by making use of a recent result proposed in [77] and show that the optimization problem involved always converges when Newton's method is used. The proposed methodology is based on the contributions made in [140], [139].

In the remainder of this chapter, we first revisit some existing works on directional statistics and its applications to stochastic filtering. This is followed by an introduction to quaternion-based orientation representation and to the Bingham distribution. In the succeeding two sections, the actual contribution is presented, which is a sampling scheme for antipodally symmetric distributions on hyperspheres and a filter making use of this sampling scheme. Finally, we evaluate the entire framework discussed in this chapter.

5.2 Related Approaches

Directional statistics has been used in a variety of applications. These involve (but are not limited to) biology [8], geology [87], the estimation of wind directions [17], seasonal data in political and social sciences [33], and time pattern analysis in crimes [15]. Applications of directional statistics involve a number of different probability distributions. From a theoretical viewpoint, the wrapped normal distribution, which was used for the first time in [106], is of particular interest. It not only appears as a limit distribution, but its p.d.f. is also a solution of the heat equation on the circle. Unfortunately, it is not closed under multiplication. That is, the product of two wrapped normal densities is not again a (possibly unnormalized) wrapped normal density.

Thus, the von Mises distribution (originally proposed in [120]) is often used as an alternative on the unit circle. Products of the von Mises densities are not only again (unnormalized) von Mises densities, but the distribution also has the maximum entropy property among all circular distributions given the first circular moment. The von Mises distribution can also be generalized to higher dimensional hyperspheres. This generalization is called the von Mises-Fisher distribution. It arises naturally by considering a Gaussian distribution with a mean given by a vector of unit length. The Bingham distribution also arises by restricting a Gaussian density to unit length. However, contrary to the von Mises-Fisher case, the original Gaussian distribution has zero mean. For this reason, a Bingham density usually has two modes.

Stochastic filtering based on directional statistics is an emerging area of research. To the best of the authors' knowledge, the first filters using directional statistics date back to the 1970s, e.g. the filter proposed in [86]. Later, a filter based on the von Mises-Fisher distribution with an identity system model was derived in [18]. More recently, Azmani [5] proposed a filter based on the von Mises distribution that also assumed an identity system model and direct measurements. A filter considering more complicated models using deterministic sampling of three samples was proposed in [143]. This sampling scheme was extended to the use of five samples in [144].

An application of the Bingham distribution is discussed in [37]. A filter was proposed and independently developed in [149], [36], [35]. These filters either assume an identity system model or, as done in [36], propose an orientational equivalent to a constant velocity model. In that work, random sampling is used to approximately propagate an uncertain quantity defined on the hypersphere. In the following, we describe a methodology for avoiding the use of random sampling in that case.

Handling of the Bingham normalization constant, which is a hypergeometric function of matrix argument [45], [94], is discussed in several works.

A very general method for computing hypergeometric functions of matrix argument was proposed in [66]. In [36], precomputed lookup tables are used for handling the normalization constant and performing parameter estimation. Use of holonomic gradient descent is proposed in [95], [69].

5.3 Quaternion-Based Orientation Representation

Quaternions were introduced by William Hamilton and are a convenient way of representing orientations in three dimensional space. In this section, we will give a brief overview of operations on quaternions and how they can be used for orientation representation. A more thorough discussion can be found in [73].

A quaternion can be thought of as a generalization of a complex number. In addition to the complex unit i, it involves two further units j and k, all of which do not commute when multiplied with one of the others. However, they are connected by the property

$$i^2 = j^2 = k^2 = ijk = -1 \ . \tag{5.1}$$

Thus, a quaternion is given using a tuple of 4 real numbers as

$$a_1 + a_2\, i + a_3\, j + a_4\, k \ ,$$

which will be denoted by a four dimensional vector in this work ($\underline{a}$ in this case).

5.3.1 Fundamental Properties

Summation of quaternions corresponds to addition of two vectors in $\mathbb{R}^4$. Multiplication of quaternions, in this work denoted by $\oplus$, is a generalization of multiplication of complex numbers and can be obtained from (5.1) as

$$\begin{pmatrix} a_1 \\ a_2 \\ a_3 \\ a_4 \end{pmatrix} \oplus \begin{pmatrix} b_1 \\ b_2 \\ b_3 \\ b_4 \end{pmatrix} := \begin{pmatrix} a_1 b_1 - a_2 b_2 - a_3 b_3 - a_4 b_4 \\ a_1 b_2 + a_2 b_1 + a_3 b_4 - a_4 b_3 \\ a_1 b_3 - a_2 b_4 + a_3 b_1 + a_4 b_2 \\ a_1 b_4 + a_2 b_3 - a_3 b_2 + a_4 b_1 \end{pmatrix} .$$

The product of quaternions is also known as the Hamilton product in reference to its inventor. Together with these two operations quaternions form a skew field, which is usually denoted by $\mathbb{H}$. Another useful operation is the conjugation of a quaternion $\underline{a}$, which will be denoted by $\underline{a}^*$. Once again, this is a generalization of the conjugation of complex numbers. It is given by

$$\underline{a}^* = (a_1, -a_2, -a_3, -a_4)^\top \ .$$

Using this definition, we can define a norm as $\sqrt{\underline{a}^* \oplus \underline{a}}$ and it can be shown that

$$\sqrt{\underline{a}^* \oplus \underline{a}} = \sqrt{a_1^2 + a_2^2 + a_3^2 + a_4^2} \ .$$

That is, the quaternion norm corresponds to the Euclidean norm $||\underline{a}||$. The inverse of a quaternion, denoted as $\underline{a}^{-1}$, is given by

$$\underline{a}^{-1} = \frac{\underline{a}^*}{||\underline{a}||^2} \ .$$

This inverse is unique even though multiplication is not commutative. The power of a quaternion $\underline{a}$ with a real exponent $c \in \mathbb{R}$ is defined as

$$\underline{a}^c = ||\underline{a}||^c \left(\begin{pmatrix} \cos(c \cdot \alpha) \\ 0 \\ 0 \\ 0 \end{pmatrix} + \frac{\underline{b}}{||\underline{b}||} \sin(c \cdot \alpha) \right) ,$$

where $\underline{b} = (0, a_2, a_3, a_4)^\top$ and α is obtained as the solution of

$$\cos(\alpha) = \frac{a_1}{||\underline{a}||} \ , \qquad \sin(\alpha) = \frac{||\underline{b}||}{||\underline{a}||} \ .$$

The multiplicative group of quaternions is also a matrix group. Consequently, every quaternion $\underline{a}$ can be represented as a real matrix

$$\mathbf{Q}_a = \begin{pmatrix} a_1 & a_2 & a_3 & a_4 \\ -a_2 & a_1 & -a_4 & a_3 \\ -a_3 & a_4 & a_1 & -a_2 \\ -a_4 & -a_3 & a_2 & a_1 \end{pmatrix} \tag{5.2}$$

and consequently $\mathbf{Q}_a^\top$ is the matrix representation of $\underline{a}^*$. Using this matrix multiplication, we can rewrite a Hamilton product of $\underline{a}$ and $\underline{b}$ in the following way

$$\underline{a} \oplus \underline{b} = \mathbf{D} \cdot \mathbf{Q}_a \cdot \mathbf{D} \cdot \underline{b} \, , \tag{5.3}$$

where $\mathbf{Q}_a$ is as defined above and $\mathbf{D} = \mathrm{diag}(1, -1, -1, -1)$. That is, the Hamilton product can also be expressed as a matrix vector multiplication. This result is more important for simplifying some of the proofs later rather than for actual computation. Similarly, we can obtain the quaternion inverse as

$$\underline{a}^{-1} = \frac{\mathbf{D} \cdot \underline{a}}{||\underline{a}||^2} \, . \tag{5.4}$$

5.3.2 Orientation Representation

Using unit quaternions for representing rotations and applying these rotations to vectors in $\mathbb{R}^3$ is very popular because operations, such as the subsequent application of rotations, can be easily implemented as a sequence of quaternion multiplications. A quaternion is used to represent a rotation with an angle α around the axis $\underline{x} \in \mathbb{R}^3$ (with $||\underline{x}|| = 1$) as follows

$$\cos\left(\frac{\alpha}{2}\right) + (x_1\, i + x_2\, j + x_3\, k) \sin\left(\frac{\alpha}{2}\right) \, .$$

From the $||\underline{x}|| = 1$ condition, it follows that this choice results in a unit quaternion. On the other hand, taking an arbitrary unit quaternion and solving it for $\underline{x}$ and θ gives the original rotation parameters. So far, we only have a rotation. However, once a reference orientation is given, this rotation representation can also be interpreted as an orientation representation. Quaternion multiplication is used to apply a rotation represented by a unit quaternion $\underline{q}$ to some vector $\underline{v} \in \mathbb{R}^3$. This is done by interpreting $\underline{v}$ as a quaternion $v_1\, i + v_2\, j + v_3\, k$ and then obtaining the newly rotated vector $\underline{\tilde{v}}$ as follows

$$\begin{pmatrix} 0 \\ \tilde{v}_1 \\ \tilde{v}_2 \\ \tilde{v}_3 \end{pmatrix} := \underline{q} \oplus \begin{pmatrix} 0 \\ v_1 \\ v_2 \\ v_3 \end{pmatrix} \oplus \underline{q}^*$$

It can be shown that the real part remains zero after these multiplications. Thus, the notation used here assigning 0 to the real part of the quaternion containing $\underline{\tilde{v}}$ is well-justified. From this representation, it can also be seen that the multiplicative inverse of a unit quaternion inverts the rotation represented by this unit quaternion. Another consequence is the fact that unit quaternions $\underline{q}$ and $-\underline{q}$ represent the same orientation. Because of this property, the group of unit dual quaternions is a so called double cover of the group of orientations in $\mathbb{R}^3$. The former is known as special unitary group of degree 2 (written as $SU(2)$), whereas the latter is known as special orthogonal group of dimension 3 (written as $SO(3)$).

The properties discussed above give rise to two further arguments for using quaternions for representing orientations. First, if for two quaternions $\underline{q}_1 \neq \pm\underline{q}_2$ then $\underline{q}_1$ and $\underline{q}_2$ represent different orientations. Thus, quaternions do not suffer from Gimbal lock, which happens for Euler angles. Second, quaternions are easier to handle in the presence of numerical cut off errors. These errors might lead to $||\underline{q}|| \neq 1$, which is easily resolved by renormalization of $\underline{q}$. When orthonormal matrices are used instead for representing orientations, numerical cut off errors result in the need to restore orthonormality. This in turn requires carrying out a much more costly Gram-Schmidt orthonormalization procedure.

For using unit quaternions in dynamic state estimation, it is necessary to define a probability distribution capable of representing uncertainty on the unit ball $S^3 \in \mathbb{R}^4$ accounting for the fact that $\underline{q}$ and $-\underline{q}$ are two different descriptions of the same orientation. Thus, we also require antipodal symmetry, that is, for the density function f the identity $f(\underline{q}) = f(-\underline{q})$ must hold.

5.4 Bingham Distribution

The Bingham distribution [9] is an antipodally symmetric distribution defined on the hypersphere $S^{n-1} \subseteq \mathbb{R}^n$ that arises naturally when conditioning a zero-mean multivariate Gaussian distribution to unit length. From this relationship, the density can be easily derived using Bayes' theorem. Let $\underline{x} \sim \mathcal{N}(\underline{0}, \mathbf{C})$, then we have

$$f\left(\underline{x} \mid ||\underline{x}|| = 1\right) \propto \delta(1 - ||\underline{x}||) \cdot f(\underline{x}) \ .$$

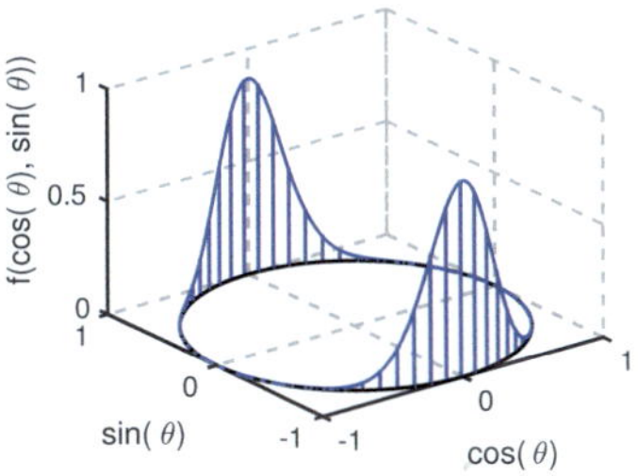

(a) $\mathbf{M} = \begin{pmatrix} 1 & 0 \\ 0 & 1 \end{pmatrix}$, $\mathbf{Z} = \text{diag}(-10, 0)$.

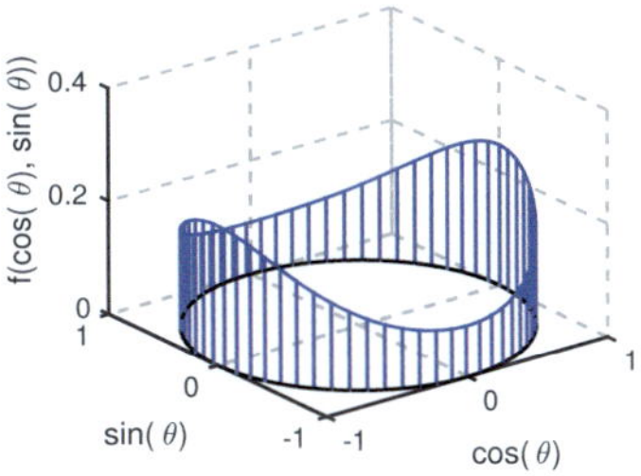

(b) $\mathbf{M} = \begin{pmatrix} 0 & 1 \\ 1 & 0 \end{pmatrix}$, $\mathbf{Z} = \text{diag}(-1, 0)$.

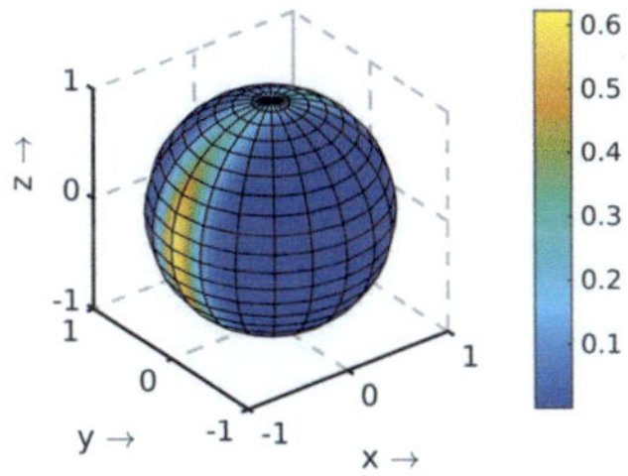

(c) $\mathbf{M} = \begin{pmatrix} 0 & 0 & 1 \\ 1 & 0 & 0 \\ 0 & 1 & 0 \end{pmatrix}$, $\mathbf{Z} = \text{diag}(-20, -1, 0)$.

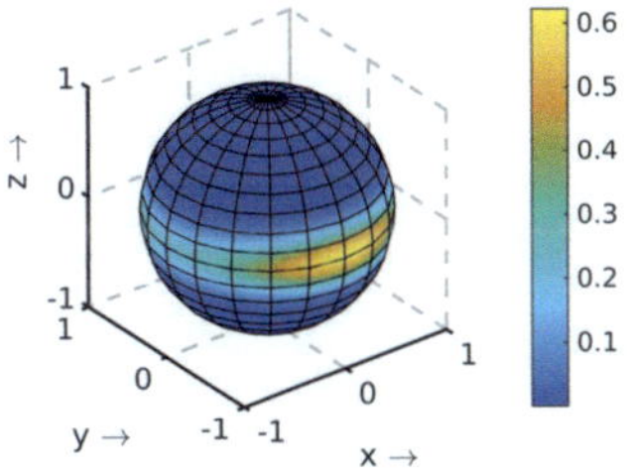

(d) $\mathbf{M} = \begin{pmatrix} 0 & 1 & 0 \\ 0 & 0 & 1 \\ 1 & 0 & 0 \end{pmatrix}$, $\mathbf{Z} = \text{diag}(-20, -1, 0)$.

Figure 5.1.: Bingham densities for the 2D and 3D case.

Thus, the density of a Bingham distribution is proportional to a Gaussian and thus given by

$$f(\underline{x}) = \frac{1}{N(-\mathbf{C}^{-1}/2)} \exp\left(-\frac{1}{2} \underline{x}^\top \mathbf{C}^{-1} \underline{x} \right) , \qquad \underline{x} \in S^{n-1} .$$

Here $N(-\mathbf{C}^{-1}/2)$ denotes the normalization constant ensuring the integral of f to be 1. Here and in what follows, integration on hyperspheres is understood as described in appendix A.2. Usually, $-\mathbf{C}^{-1}/2$ is replaced by its eigendecomposition $\mathbf{M}\,\mathbf{Z}\,\mathbf{M}^\top$. Thus, $\mathbf{M}$ is an orthonormal matrix, where each column consists of normalized eigenvectors of $-\mathbf{C}^{-1}/2$ and $\mathbf{Z}$ is a diagonal matrix consisting of the corresponding eigenvalues. According to the transformation theorem and as a consequence of the fact that $|\det(\mathbf{M})| = 1$, the computation of the normalization constant can be simplified in the following way

$$\begin{aligned}
N(\mathbf{M}\,\mathbf{Z}\,\mathbf{M}^\top) &= \int_{S^{n-1}} \exp\left(\underline{x}^\top \mathbf{M}\,\mathbf{Z}\,\mathbf{M}^\top \underline{x}\right)\, \mathrm{d}\underline{x} \\
&= \int_{S^{n-1}} \exp\left(\underline{x}^\top \mathbf{Z}\,\underline{x}\right)\, \mathrm{d}\underline{x} \\
&= N(\mathbf{Z}) \ .
\end{aligned}$$

Thus, the following formulation of the density of a Bingham distribution is justified

$$f(\underline{x}) = \frac{1}{N(\mathbf{Z})} \exp\left(\underline{x}^\top \mathbf{M}\,\mathbf{Z}\,\mathbf{M}^\top \underline{x}\right) \ , \qquad \underline{x} \in S^{n-1} \ .$$

For a random variable $\underline{x}$ distributed according to the Bingham distribution with parameters $\mathbf{M}$ and $\mathbf{Z}$, we will use the notation $\underline{x} \sim \text{Bingham}(\mathbf{M}, \mathbf{Z})$. For better interpretation, the entries in $\mathbf{Z}$ are usually given in an ascending order. This is not a restriction on the choice of possible densities because this ordering can easily be obtained by a corresponding resorting of columns in $\mathbf{M}$ from every possible original combination of parameter matrices $\mathbf{M}$ and $\mathbf{Z}$. After reordering, the last column $\underline{m}_n$ of $\mathbf{M}$ can be interpreted as the mode of the Bingham distribution. This is easily seen from the fact that $\underline{m}_n$ is the eigenvector corresponding to the largest eigenvalue of $\mathbf{M}\,\mathbf{Z}\,\mathbf{M}^\top$ and thus $\exp\left(\underline{x}^\top \mathbf{M}\,\mathbf{Z}\,\mathbf{M}^\top \underline{x}\right)$ is maximized for $\underline{x} = \underline{m}_n$. A similar argument shows that the remaining columns of $\mathbf{M}$ describe the directions of the main axes. The matrix $\mathbf{Z}$ does not need to be negative definite. This is because the Bingham distribution is defined on a compact domain and can also be seen from the following relationship. For all $c \in \mathbb{R}$, it holds

$$\begin{aligned}
N(\mathbf{Z} + c\,\mathbf{I}) &= \int_{S^{n-1}} \exp\left(\underline{x}^\top (\mathbf{Z} + c\,\mathbf{I})\, \underline{x}\right)\, \mathrm{d}\underline{x} \\
&= \int_{S^{n-1}} \exp\left(\underline{x}^\top \mathbf{Z}\,\underline{x} + \underline{x}^\top (c\,\mathbf{I})\, \underline{x}\right)\, \mathrm{d}\underline{x}
\end{aligned}$$

$$= \exp(c) \cdot \int_{S^{n-1}} \exp\left(\underline{x}^\top \mathbf{Z}\, \underline{x}\right)\, \mathrm{d}\underline{x}$$
$$= \exp(c) \cdot N(\mathbf{Z})\ .$$

Several examples of the Bingham density are visualized in Figure 5.1.

In order to obtain a unique representation of the distribution, we will usually expect the last entry of $\mathbf{Z}$ to be 0. Because of antipodal symmetry, the first moment of a Bingham$(\mathbf{M}, \mathbf{Z})$ distributed random vector $\underline{x}$ is given by $\mathbb{E}(\underline{x}) = \underline{0}$. The second moment is more complicated and was derived in Binghams original work [9] as

$$\mathbb{E}\left(\underline{x} \cdot \underline{x}^\top\right) = \mathbf{M} \cdot \operatorname{diag}\left(e_1, \ldots, e_n\right) \cdot \mathbf{M}^\top\ , \tag{5.5}$$

with

$$e_i = \left.\frac{\partial N(\mathbf{A})}{\partial a_i}\right|_{\mathbf{A}=\mathbf{Z}} \cdot N(\mathbf{Z})^{-1}$$

and $\mathbf{A} = \operatorname{diag}(a_1, ..., a_n)$. A useful property of the Bingham distribution is the fact that it is the maximum entropy distribution on the hypersphere given the second moment and $\mathbb{E}(\underline{x}) = 0$, which is a hyperspherical analogue to a similar property of the Gaussian. This result was originally derived in [88] and extends an earlier result from [57]. Another property which the Bingham distribution has in common with a Gaussian, is the fact that the product of two Bingham densities is itself another (unnormalized) Bingham density. The proof is carried out analogously to the proof of the corresponding statement for Gaussian densities. Let f_i be the density of Bingham$(\mathbf{M}_i, \mathbf{Z}_i)$ for $i = 1, 2$. Then

$$\begin{aligned} f_1(\underline{x}) \cdot f_2(\underline{x}) &\propto \exp\left(\underline{x}^\top \mathbf{M}_1\, \mathbf{Z}_1\, \mathbf{M}_1^\top\, \underline{x} + \underline{x}^\top \mathbf{M}_2\, \mathbf{Z}_2\, \mathbf{M}_2^\top\, \underline{x}\right) \\ &= \exp\left(\underline{x}^\top \mathbf{M}\, \mathbf{Z}\, \mathbf{M}^\top\, \underline{x}\right)\ , \end{aligned} \tag{5.6}$$

where $\mathbf{M}\,\mathbf{Z}\,\mathbf{M}^\top$ is the eigendecomposition of $\mathbf{M}_1\, \mathbf{Z}_1\, \mathbf{M}_1^\top + \mathbf{M}_2\, \mathbf{Z}_2\, \mathbf{M}_2^\top$. As will be seen later, this property is of particular use because it makes a measurement update in closed form possible.

Example 5.1. *Contrary to the Gaussian case, multiplication of two Bingham densities and subsequent renormalization might result in a Bingham density with higher uncertainty than one of the original densities. This* *is seen by considering two three-dimensional Bingham distributions. The first is given by* $\mathbf{Z}_1 = \mathrm{diag}(-10, -10, 0)$ *and* $\mathbf{M}_1 = \mathbf{I}$ *and the second is given by* $\mathbf{Z}_2 = \mathrm{diag}(-50, 0, 0)$ *and*

$$\mathbf{M}_2 = \begin{pmatrix} 0 & 0 & 1 \\ 0 & 1 & 0 \\ 1 & 0 & 0 \end{pmatrix}$$

Obviously, we have $\mathbf{M}_1\,\mathbf{Z}_1\,\mathbf{M}_1^\top = \mathbf{Z}_1$ *and* $\mathbf{M}_2\,\mathbf{Z}_2\,\mathbf{M}_2^\top = \mathrm{diag}(0, 0, -50)$. *Multiplying the respective densities as in* (5.6) *yields an unnormalized Bingham density.*

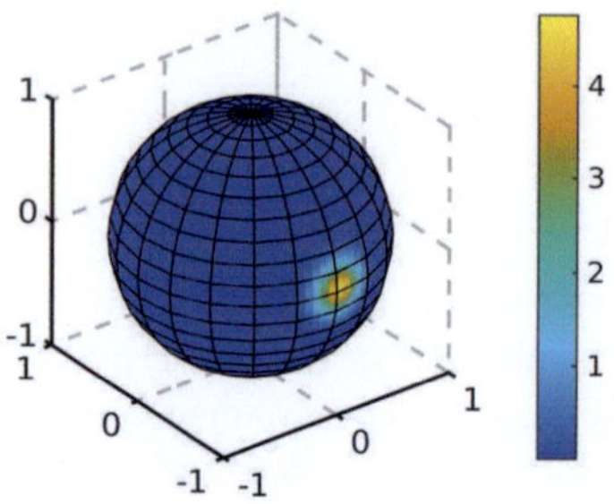

(a) Bingham($\mathbf{M}_1, \mathbf{Z}_1$) density.

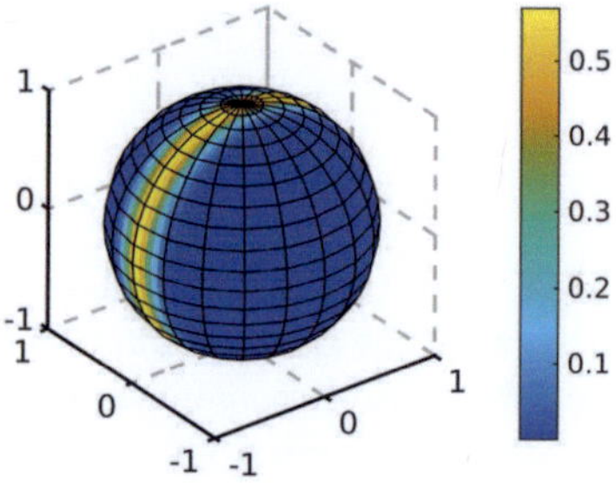

(b) Bingham($\mathbf{M}_2, \mathbf{Z}_2$) density.

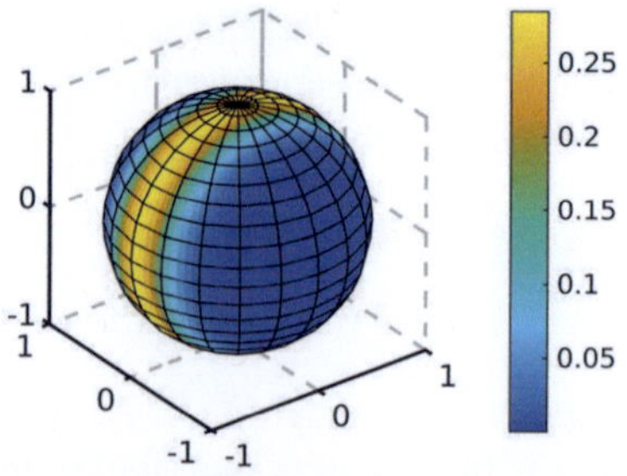

(c) Renormalized product.

Figure 5.2.: Visualization of Example 5.1 showing that the renormalized product of two Bingham densities may yield a Bingham density with higher uncertainty, which is not possible for the product of two Gaussian densities.

The distribution parameters $\mathbf{M}$ *and* $\mathbf{Z}$ *are obtained from* $\mathbf{M}_1\,\mathbf{Z}_1\,\mathbf{M}_1^\top + \mathbf{M}_2\,\mathbf{Z}_2\,\mathbf{M}_2^\top = \text{diag}(-10,-10,-50)$. *Performing an eigendecomposition and reshifting the entries of* $\mathbf{Z}$ *such that its largest entry is zero yields* $\mathbf{M} = \mathbf{M}_2$ *and* $\mathbf{Z} = \text{diag}(-40,0,0)$. *The corresponding densities are visualized in Figure 5.2.*

This example has an important consequence for working with the Bingham distribution. In the filtering application considered in this chapter, it might happen that uncertainty increases after a measurement update. A simple interpretation of this effect is that a better estimate is not necessarily one with less uncertainty. However, this effect may also happen for nonlinear filtering in linear state spaces.

5.4.1 Parameter Estimation

The proposed parameter estimation procedure is based on matching the second moment (5.5). Thus, given a number of random samples $\underline{s}_1, ..., \underline{s}_m$, our goal is to find orthonormal $\mathbf{M}$ and diagonal $\mathbf{Z}$ such that for $\underline{x} \sim \text{Bingham}(\mathbf{M}, \mathbf{Z})$ we have

$$\mathbb{E}\left(\underline{x}\cdot\underline{x}^\top\right) = \underbrace{\frac{1}{m}\sum_{i=1}^{m}\underline{s}_i\cdot\underline{s}_i^\top}_{:=\boldsymbol{\Sigma}} \ . \tag{5.7}$$

This is where our restriction $z_n = 0$ and the ascending ordering of the z_i become useful. Without both conditions, the above equation never has a unique solution when solved for $\mathbf{M}$ and $\mathbf{Z}$. For many operations, this uniqueness is not necessary. However, it simplifies both, the interpretation of the density and the actual computation. In Bingham's original work, it has been shown that this approach is not only the parameter estimation procedure in the sense of the method of moments, but also provides the maximum likelihood estimate of the distribution parameters.

The estimation itself is a numerical procedure that consists of two steps. First, we obtain the estimate $\hat{\mathbf{M}}$ as the orthonormal matrix of eigenvectors of $\mathbf{\Sigma}$. Second, we obtain the estimate $\hat{\mathbf{Z}}$ by solving the system of equations

$$c_i = \left.\frac{\partial N(\mathbf{Z})}{\partial z_i}\right|_{\mathbf{Z}=\hat{\mathbf{Z}}} \cdot N(\hat{\mathbf{Z}})^{-1} \tag{5.8}$$

for $\hat{z}_i$ with $\hat{\mathbf{Z}} = \text{diag}(\hat{z}_1, \dots, \hat{z}_n)$ under the condition that $\hat{z}_n = 0$. That is, we actually have to solve only using $n-1$ unknown variables. To avoid numerical problems, it is necessary to ensure that $z_i \leq z_{i+1}$. The entire procedure is shown in algorithm 5.1.

Algorithm 5.1 Bingham Parameter Estimation

procedure ESTIMATEBINGHAM($\underline{s}_1, \dots, \underline{s}_m$)
  $\mathbf{C} \leftarrow \frac{1}{m}\sum_{i=1}^{m} \underline{s}_i \cdot \underline{s}_i^\top$;
  $(\underline{v}_1, \dots, \underline{v}_n, c_1, \dots, c_n) \leftarrow \text{Eigendecomposition}(\mathbf{C})$;
  $(\hat{z}_1, \dots, \hat{z}_{n-1}) \leftarrow \text{Solve}\left((5.8) \text{ with } \tilde{\mathbf{Z}} = \text{diag}(z_1, ..., z_{n-1}, 0)\right)$;
  $\hat{z}_n \leftarrow 0$;
  $(i_1, \dots, i_n) \leftarrow \text{GetOrderAsc}(\hat{z}_1, \dots, \hat{z}_n)$; ▷ Correct order of entries in $\hat{\mathbf{Z}}$
  $\hat{\mathbf{Z}} \leftarrow \text{diag}(\hat{z}_{i_1} - \hat{z}_{i_n}, ..., \hat{z}_{i_{n-1}} - \hat{z}_{i_n}, 0)$; ▷ Construct $\hat{\mathbf{Z}}$ according to this order.
  $\hat{\mathbf{M}} \leftarrow (\underline{v}_{i_1}, \dots, \underline{v}_{i_n})$; ▷ Construct $\hat{\mathbf{M}}$ according to this order.
  return $(\hat{\mathbf{M}}, \hat{\mathbf{Z}})$;
end procedure

This algorithm can also be formulated in a way where each sample is assigned a weight. This might be of interest when using the Bingham distribution in the context of particle filtering. Furthermore, we will use the notation EstimateBingham($\mathbf{C}$) whenever the second moment matrix is already given. In that case the algorithm stays the same, except for the first step (where $\mathbf{C}$ is computed), which can be left out.

5.4.2 Computation of the Normalization Constant and its Derivatives

In the entire estimation procedure and, as will be seen, in our filtering algorithm, computation of the normalization constant plays a crucial role.

This is also true for computation of its derivatives. A naïve numerical integration approach for computing the normalization constant would result in the need for multidimensional integration. This seems infeasible in most applications with real-time requirements. Thus, we need a more thorough investigation of the challenge of computing the normalization constant.

First, we observe that the normalization constant can be expressed in terms of a hypergeometric function of matrix argument

$$N(\mathbf{Z}) = |S^{n-1}| \cdot {}_1F_1\left(\frac{1}{2}; \frac{n}{2}; \mathbf{Z}\right) ,$$

where $|S^{n-1}|$ denotes the surface area of the hypersphere, i.e.,

$$|S^{n-1}| = \frac{n\,\pi^{n/2}}{\Gamma\left(\frac{n}{2}+1\right)} .$$

A formal definition and a discussion of hypergeometric functions of matrix argument ${}_pF_q$ can be found in [66], [94], [45]. However, there is currently no general method for computing hypergeometric functions of matrix argument with sufficient precision at feasible computational cost. So far, the best general algorithm seems to be [66].

Thus, we use an approach based on saddlepoint approximations [22], [40]. These were used to approximate the Bingham normalization constant in [77], which we also use in what follows. The third order saddlepoint approximation is of particular interest, because it outperforms the other proposed approximations in terms of precision. It is given by

$$\tilde{N}(\mathbf{Z}) := \sqrt{\frac{2\,\pi^{(n-1)}}{K^{(2)}(\hat{t},\mathbf{Z})}} \cdot \left(\prod_{i=0}^{n} \left(\sqrt{-z_i} - \hat{t}\right)^{-1/2}\right) \cdot \exp(-\hat{t} + T) ,$$

where

$$T := \frac{\rho_4(\hat{t},\mathbf{Z})}{8} - \frac{5\,\rho_3(\hat{t},\mathbf{Z})}{24} ,$$

$$\rho_j(t,\mathbf{Z}) := \frac{K^{(j)}(\hat{t},\mathbf{Z})}{K^{(2)}(\hat{t},\mathbf{Z})^{j/2}} ,$$

$$K^{(j)}(\hat{t}, \mathbf{Z}) := \sum_{i=1}^{n} \left(\frac{(j-1)!}{2} \cdot \frac{1}{(\sqrt{-z_i} - t)^j} \right) .$$

Furthermore, $\hat{t}$ is obtained by solving $K^{(1)}(t, \mathbf{Z}) = 1$ for t on $(-\infty, z^*)$, where $z^* = \min_i(\sqrt{-z_i})$. The following proposition guarantees an efficient computation of $\tilde{N}(\mathbf{Z})$ by establishing that solving $K^{(1)}(t, \mathbf{Z}) = 1$ can be done efficiently.

Proposition 5.2. *The function $K^{(1)}(t, \mathbf{Z})$ is convex w.r.t. t on $(-\infty, z^*)$.*

Proof. This is easily seen from the second derivative of $K^{(1)}$, which is given by

$$\frac{\partial^2 K^{(1)}(t, \mathbf{Z})}{\partial t^2} = \sum_{i=1}^{n} \frac{1}{(\sqrt{-z_i} - t)^3}$$

Each term is positive on the considered interval, and thus, the convexity follows from the positivity of the entire second derivative. □

Now, Newton's method can be used efficiently for solving $K^{(1)}(t, \mathbf{Z}) = 1$ for t. The presented approximations can also be used to obtain approximated derivatives. The obvious way to do this would be the use of finite-differences. Here, we take an approach based on a result proposed in [76]. This result establishes a relation between the derivatives of the Bingham Normalization constant and a normalization constant for a higher dimensional Bingham distribution.

Corollary 5.3. *The derivative of the Bingham normalization constant with respect to z_i is given by*

$$\frac{\partial N(\mathbf{Z})}{\partial z_i} = \frac{1}{2\pi} N(\mathrm{diag}(z_1, z_2, \ldots, z_{i-1}, z_i, z_i, z_i, z_{i+1}, z_{i+2}, \ldots, z_n)) .$$

Proof. This is an immediate consequence of Proposition 1 from [76]. □

5.4.3 Representation of Uncertain Orientations

So far, it is easily seen that the four dimensional Bingham distribution can be used for describing uncertainty on the domain of orientations by interpreting it as a probability distribution over the quaternion unit sphere. Its antipodal symmetry ensures a correct consideration of the fact that $\underline{a}$ and $-\underline{a}$ represent the same orientation.

The mode of the Bingham distribution can be thought of as the orientational mean. It also corresponds to the hyperspherical mean estimator based on the eigendecomposition of a scatter matrix. Thus, a Bingham distribution with a certain desired mean can be constructed by deliberately choosing the last column of $\mathbf{M}$. A Bingham distribution where the last entry of $\mathbf{M}$ equals to $(1,0,0,0)^\top$ can be thought of as a zero mean equivalent for uncertain orientations. The remaining columns need to be chosen in a way ensuring $\mathbf{M}$ to be orthonormal, e.g.,

$$\mathbf{M} = \begin{pmatrix} 0 & 0 & 0 & 1 \\ 0 & 0 & 1 & 0 \\ 0 & 1 & 0 & 0 \\ 1 & 0 & 0 & 0 \end{pmatrix}$$

However, the choice of the remaining columns of $\mathbf{M}$ is not only restricted by the need for orthonormality. These columns also have a probabilistic interpretation, because they represent the principal axes of the uncertainty of the Bingham distribution, and thus, have an impact on which orientations are more likely than others. This involves the representation of a uniform distribution by choosing $\mathbf{Z} = \mathrm{diag}(0,0,0,0)$ (in that case the choice of $\mathbf{M}$ is obviously arbitrary) or distributions with circular symmetry (that is, the p.d.f. has a constant value on an entire axis), e.g., by choosing $\mathbf{Z} = \mathrm{diag}(-10,-10,0,0)$

5.5 A Deterministic Sampling Scheme

The deterministic sampling method proposed in this section works for arbitrary hyperspherical and antipodally symmetric probability densities. For approximating the second moment of a distribution (or samples) on S^{n-1}, it places $4n-2$ samples such that these have the same second moment. Thus, it can be thought of as a deterministic sampling method for placing antipodally symmetric samples on the unit sphere in a way such that a predefined second moment matrix is satisfied. The only assumption we make is that this second moment matrix stems from samples of a distribution defined on the hypersphere.

The proposed method is subdivided into two steps. First, we derive the sampling scheme for the case of diagonal $E(\underline{x}\,\underline{x}^\top) = \mathrm{diag}(c_1, \ldots, c_n)$ with non-decreasing entries, i.e. $c_i \leq c_{i+1}$. Second, we generalize this to arbitrary second moment matrices using their respective eigendecompositions.

Let $\mathbf{C} = \mathrm{diag}(c_1, ..., c_n)$ be the diagonal matrix representing the second moment that needs to be approximated. We also assume that our first goal is obtaining a discrete distribution defined on the hypersphere with $\mathbb{E}(\underline{x}\,\underline{x}^\top) = \mathbf{C}$. Due to the fact that $\mathbf{C}$ is positive definite, all of its diagonal entries are positive or 0. Furthermore, it can be observed that all diagonal entries of $\mathbf{C}$ sum up to 1. The latter is seen using the fact that the trace operator only performs linear operations and thus

$$\mathrm{tr}(\mathbf{C}) = \mathrm{tr}(\mathbb{E}(\underline{x}\,\underline{x}^\top)) = \mathbb{E}(\mathrm{tr}(\underline{x}\,\underline{x}^\top)) = \mathbb{E}(1) = 1 \; .$$

We place our initial sample $\underline{x}_1$ at the pole $(0, \ldots, 0, 1)^\top$. The next samples are placed by choosing main axes on the hypersphere around that pole. Usually, $n-1$ scalar values are needed in order to describe the position of a point on an n-dimensional hypersphere. Thus, $2n-2$ samples are placed on the hypersphere around the pole in the style of the UKF. The samples $\underline{x}^{(2)}$ to $\underline{x}^{(2n-1)}$ are generated in the following manner. The i-th entry of the Samples $2i$ and $2i+1$ is given by

$$\begin{aligned} \underline{x}_{2i}^{(i)} &= \sin(\alpha_i) \; , \\ \underline{x}_{2i+1}^{(i)} &= -\sin(\alpha_i) \end{aligned}$$

with a suitably chosen α_i. This choice is discussed below. The last entry of both samples is given by

$$\underline{x}_{2i}^{(n)} = \underline{x}_{2i+1}^{(n)} = \cos(\alpha_i) \; .$$

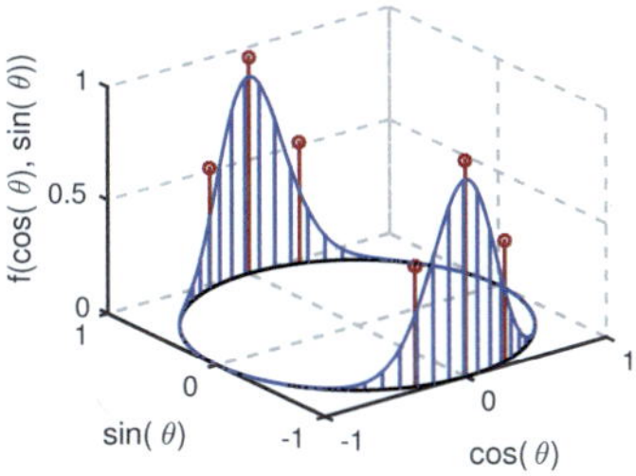

(a) $\mathbf{M} = \begin{pmatrix} 1 & 0 \\ 0 & 1 \end{pmatrix}$, $\mathbf{Z} = \mathrm{diag}(-10, 1)$.

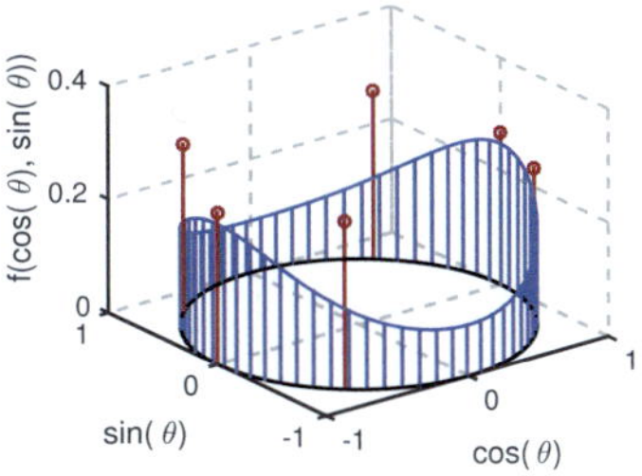

(b) $\mathbf{M} = \begin{pmatrix} 0 & 1 \\ 1 & 0 \end{pmatrix}$, $\mathbf{Z} = \mathrm{diag}(-1, 0)$.

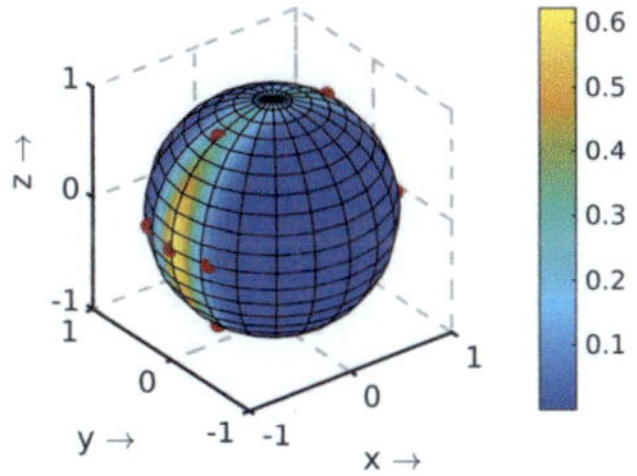

(c) $\mathbf{M} = \begin{pmatrix} 0 & 0 & 1 \\ 1 & 0 & 0 \\ 0 & 1 & 0 \end{pmatrix}$, $\mathbf{Z} = \mathrm{diag}(-20, -1, 0)$.

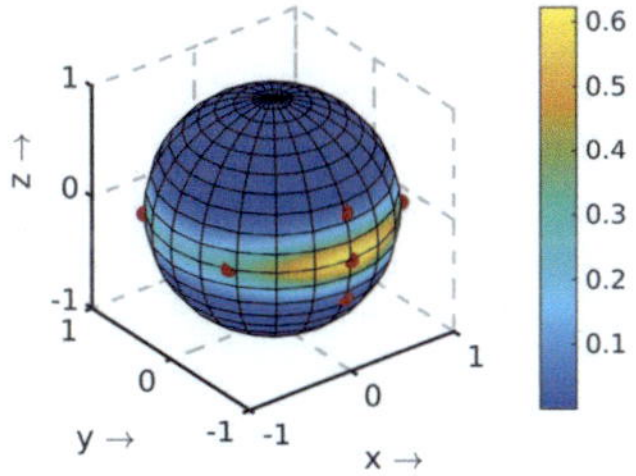

(d) $\mathbf{M} = \begin{pmatrix} 0 & 1 & 0 \\ 0 & 0 & 1 \\ 1 & 0 & 0 \end{pmatrix}$, $\mathbf{Z} = \mathrm{diag}(-20, -1, 0)$.

Figure 5.3.: Examples for Bingham densities with deterministic sampling using sampling parameter $\lambda = 0.5$.

All other entries are set to zero. The first samples generated in this manner are

$$\begin{aligned} \underline{x}_2 &= (\sin(\alpha_1), 0, \ldots, 0, \cos(\alpha_1))^\top , \\ \underline{x}_3 &= (-\sin(\alpha_1), 0 \ldots, 0, \cos(\alpha_1))^\top , \\ \underline{x}_4 &= (0, \sin(\alpha_2), 0 \ldots, 0, \cos(\alpha_2))^\top , \\ \underline{x}_5 &= (0, -\sin(\alpha_2), 0 \ldots, 0, \cos(\alpha_2))^\top , \\ &\vdots \end{aligned}$$

So far, $2n-1$ samples have been generated. In order to account for antipodal symmetry, a second pole needs to be considered. This is done by negation of the existing $2n-1$ samples

$$\underline{x}_{2n} = -\underline{x}_1, \quad \cdots \quad \underline{x}_{4n-2} = -\underline{x}_{2n-1} .$$

A probability needs to be assigned to each sample for generating a discrete probability distribution. For each of the poles, this probability shall be denoted by $p_1 = p_{2n} = m_n/2$. Each of the samples x_{2i}, x_{2i+1}, $x_{2(n+i)-1}$, and $x_{2(n+i)}$ shall have probability mass denoted by $p_{2i} = p_{2i+1} = p_{2(n+i)-1} = m_i/4$ (for $i = 1, \ldots, n-1$). This introduces the conditions $m_i \geq 0$ and

$$\sum_{i=1}^{n} m_i = 1 . \tag{5.9}$$

In the next step, the actual moment matching procedure is carried out. That is, angles α_i and weights m_i need to be found so that the resulting deterministic sampling set matches the predefined second moment.

$$\begin{aligned}
\mathbf{C} = \operatorname{diag}(c_1, ..., c_n) &\stackrel{!}{=} \mathbb{E}\left(\underline{x} \cdot \underline{x}^\top\right) \\
&= \sum_{i=1}^{4n-2} p_i \cdot \underline{x}^{(i)} \, (\underline{x}^{(i)})^T \\
&= 2 \cdot \sum_{i=1}^{2n-1} p_i \cdot \underline{x}^{(i)} \, (\underline{x}^{(i)})^T \\
&= \operatorname{diag}\Bigg(m_1 \sin(\alpha_1)^2, \; \ldots, \; m_{n-1} \sin(\alpha_{n-1})^2, \\
&\qquad\qquad m_n + \sum_{i=1}^{n-1} m_i \cos(\alpha_i)^2 \Bigg) .
\end{aligned}$$

This is matched with the predefined second moment, i.e., the covariance matrix in the case of our Bingham distribution. Matching the second moment with the Bingham covariance matrix is justified, because for all antipodally symmetric distributions the mean is 0, and thus, the covariance matrix corresponds to the second moment. Solving the above equation for the angles α_i gives

$$\alpha_i = \arcsin\left(\sqrt{c_i/m_i}\right) .$$

We choose $m_n = \lambda \cdot c_n$ and $m_i = c_i + (1-\lambda)c_n/(n-1)$ for all $i = 1, \ldots, n-1$ and with $\lambda \in [0, 1)$. Using the fact that $\mathrm{tr}(\mathbf{C}) = 1$, it can be shown that (5.9) is satisfied, because

$$\sum_{i=1}^{n} m_i = \lambda \cdot c_n + \sum_{i=1}^{n-1} \left(c_i + (1-\lambda)\frac{c_n}{n-1} \right) = c_n + \sum_{i=1}^{n-1} c_i = 1 .$$

The restriction of λ to $[0, 1)$ is used in order to ensure $0 \leq p_i$.

Our method, as derived so far, is capable of approximating arbitrary antipodally symmetric distributions which have diagonal second moment matrices with increasing entries. A simple application of the eigendecomposition generalizes this to arbitrary second moment matrices $\mathbf{C}$ with $\mathrm{tr}(\mathbf{C}) = 1$. First, let $\mathbf{V} \cdot \mathrm{diag}(c_1, ..., c_n) \cdot \mathbf{V}^\top$ denote the eigendecomposition of $\mathbf{C}$. Here, the c_i denote the eigenvalues of $\mathbf{C}$ and, once again, we can assume them to be given in a non-decreasing order, i.e., $c_i \leq c_{i+1}$. Then, the procedure outlined above yields a sample set consisting of samples $\underline{x}_i$ and weights p_i with second moment $\mathrm{diag}(c_1, \ldots, c_n)$. In order to obtain a sample set with the desired second moment matrix, we simply replace sample $\underline{x}_i$ by $\mathbf{V} \cdot \underline{x}_i$. This gives us the desired result

$$\begin{aligned}
\mathbb{E}\left(\mathbf{V}\underline{x}(\mathbf{V}\underline{x})^\top\right) &= \sum_{i=1}^{4n-2} p_i \cdot (\mathbf{V} \cdot \underline{x}_i) \cdot (\mathbf{V} \cdot \underline{x}_i)^\top \\
&= \mathbf{V} \cdot \left(\sum_{i=1}^{4n-2} p_i \cdot \underline{x}_i \cdot \underline{x}_i^\top \right) \cdot \mathbf{V}^\top \\
&= \mathbf{V} \cdot \mathrm{diag}(c_1, \ldots, c_n) \cdot \mathbf{V}^\top = \mathbf{C}
\end{aligned}$$

The entire resulting algorithm is shown in Algorithm 5.2. Further research is needed for an optimal choice of λ.

Algorithm 5.2 Deterministic Sampling

procedure DETERMINISTICSAMPLING($\mathbf{C}$, λ)
  $(\underline{v}_1, \ldots, \underline{v}_n, c_1, \ldots, c_n) \leftarrow \text{Eigendecomposition}(\mathbf{C})$; ▷ Assumes $c_i < c_{i+1}$.
  $\mathbf{V} \leftarrow (\underline{v}_1, ..., \underline{v}_n)$;
  $\underline{x}_1 \leftarrow \mathbf{V} \cdot (1, \ldots, 0)^\top$;
  $\underline{x}_{2n} \leftarrow -\underline{x}_1$;
  $p_1, p_{2n} \leftarrow \lambda \cdot c_n$;
  $\underline{x}_1, \ldots, \underline{x}_{2n-1} \leftarrow (0, \ldots, 0)^\top$; ▷ Initialization
  for $i \in \{1, ..., n-1\}$ **do**
    $p_{2i}, p_{2i+1}, p_{2(i+n)-1}, p_{2(i+n)} \leftarrow \frac{1}{4}\left(c_i + (1-\lambda)\frac{c_n}{n-1}\right)$;
    $\alpha_i \leftarrow \arcsin\left(\sqrt{4 \cdot c_i / w_{2i}}\right)$;
    $\underline{x}_{2i}^{(1)}, \underline{x}_{2i+1}^{(1)} \leftarrow \cos(\alpha_i)$;
    $\underline{x}_{2i}^{(i+1)} \leftarrow \sin(\alpha_i)$;
    $\underline{x}_{2i+1}^{(i+1)} \leftarrow -\sin(\alpha_i)$;
    $\underline{x}_{2i} \leftarrow \mathbf{V} \cdot \underline{x}_{2i}$;
    $\underline{x}_{2i+1} \leftarrow \mathbf{V} \cdot \underline{x}_{2i+1}$;
    $\underline{x}_{2(i+n)-1} \leftarrow -\underline{x}_{2i}$;
    $\underline{x}_{2(i+n)} \leftarrow -\underline{x}_{2i+1}$;
  end for
  return $(p_1, \underline{x}_1), ..., (p_{4n-2}, \underline{x}_{4n-2})$;
end procedure

5.6 The Unscented Bingham Filter

In the Bingham filters presented in [149], [36], the proposed filtering schemes were capable of handling system models given by

$$\underline{x}_{t+1} = \underline{x}_t \oplus \underline{a}_t \oplus \underline{w}_t \ , \tag{5.10}$$

where $\underline{x}_t \in S^3$ (or alternatively S^1) is an uncertain system state, $\underline{a}_t$ is a control input, and $\underline{w}_t$ is system noise, that is also assumed to be distributed according to a Bingham$(\mathbf{M}_t^w, \mathbf{Z}_t^w)$ distribution. This control input can also be removed from this notation leaving $\underline{x}_t \oplus \underline{w}_t$ as consideration of control inputs is possible by a suitable choice of $\mathbf{M}_t^w$ making use of the fact that for arbitrary (but fixed) unit quaternions $\underline{a}$ and Bingham distributed random vectors $\underline{w}$, the quantity $\underline{a} \oplus \underline{w}$ is again a Bingham distributed random vector. The same is true for $\underline{w} \oplus \underline{a}$.

In this work, we generalize (5.10) and make consideration of more complicated system functions possible. That is, the system dynamics are assumed to be described by

$$\underline{x}_{t+1} = g(\underline{x}_t) \oplus \underline{w}_t \ , \tag{5.11}$$

with state $\underline{x}_t$, system noise $\underline{w}_t \sim \text{Bingham}(\mathbf{M}_t^w, \mathbf{Z}_t^w)$, and a control input $\underline{a}_t$. The system function $g : S^3 \to S^3$ is assumed to respect antipodal symmetry in the sense that $g(-\underline{x}) = -g(\underline{x})$. Thus, g can also be seen as a mapping on $SO(3)$.

The conditions imposed on g could be weakened. First, the entire filter could also be adapted to work in scenarios where g does not respect antipodal symmetry. Second, more general system models, such as $\underline{x}_{t+1} = g(\underline{x}_t, \underline{w}_t)$ could also be considered in a very similar manner. However, for the particular case of orientation estimation, the imposed restrictions seem meaningful.

As a measurement model, we consider noisy direct measurements resulting in

$$\underline{z}_t = \underline{x}_t \oplus \underline{v}_t \tag{5.12}$$

with $\underline{v}_t \sim \text{Bingham}(\mathbf{M}_t^v, \mathbf{Z}_t^v)$. Once again, this can account for measurement models of the type $\underline{x}_t \oplus \underline{a}_t \oplus \underline{v}_t$ for some unit quaternion $\underline{a}_t$.

As usual, the proposed filter is subdivided in a prediction step and a measurement update. It gives the estimates of the true system state in terms of a Bingham distribution density, that is in terms of its parameters $\mathbf{M}_t^e$, $\mathbf{Z}_t^e$.

5.6.1 Prediction Step

The prediction step starts off with an estimate of $\underline{x}_t$ in terms of its Bingham distribution parameters. Unfortunately, $g(\underline{x}_t^e) \oplus \underline{w}_t$ is not again a Bingham distributed quantity even in the case where g is the identity mapping. Thus, the basic idea of our prediction step is approximating the true distribution of $g(\underline{x}_t^e) \oplus \underline{w}_t$ by a Bingham distribution that matches $\mathbf{C}_{t+1}^p := \mathbb{E}((g(\underline{x}_t^e) \oplus \underline{w}_t) \cdot (g(\underline{x}_t^e) \oplus \underline{w}_t)^\top)$.

However, computation of this covariance matrix is in general not possible in closed form. Thus, it would require a numerical integration procedure involving multi dimensional integrals. This is not feasible for many applications with limited computation time. Thus, we approximately compute the desired moments with the help of deterministic sampling. Consequently, the entire procedure involves two approximations. First, it approximates the second moment matrix of $g(\underline{x}_t^p) \oplus \underline{w}_t$. Second, it approximates the true distribution after prediction by a Bingham distribution with the second moment obtained from the first approximation.

For obtaining $\mathbf{C}_{t+1}^p$, we sample Bingham$(\mathbf{M}_t^e, \mathbf{Z}_t^e)$ deterministically and compute the covariance $\mathbf{C}_t^w$. Then, the resulting samples $\underline{s}_i$ (and their respective weights p_i) are used together with (5.2) and (5.3) to obtain

$$
\begin{aligned}
\mathbf{C}_{t+1}^p &= \mathbb{E}((g(\underline{x}_t^e) \oplus \underline{w}_t) \cdot (g(\underline{x}_t^p) \oplus \underline{w}_t)^\top) \\
&\approx \mathbb{E}\left(\sum_{i=1}^{4n-2} p_i \left((g(\underline{s}_i) \oplus \underline{w}_t) \cdot (g(\underline{s}_i) \oplus \underline{w}_t)^\top \right) \right) \\
&= \sum_{i=1}^{4n-2} p_i \, \mathbb{E}\left((g(\underline{s}_i) \oplus \underline{w}_t) \cdot (g(\underline{s}_i) \oplus \underline{w}_t)^\top \right) \\
&= \sum_{i=1}^{4n-2} p_i \, \mathbb{E}((\underbrace{\mathbf{D} \cdot \mathbf{Q}_i \cdot \mathbf{D}}_{:=\mathbf{M}} \cdot \underline{w}_t) \cdot (\mathbf{D} \cdot \mathbf{Q}_i \cdot \mathbf{D} \cdot \underline{w}_t)^\top) \\
&= \sum_{i=1}^{4n-2} p_i \cdot \mathbf{M} \cdot \mathbb{E}\left(\underline{w}_t \cdot \underline{w}_t^\top \right) \cdot \mathbf{M}^\top \\
&= \sum_{i=1}^{4n-2} p_i \cdot \mathbf{M} \cdot \mathbf{C}_t^w \cdot \mathbf{M}^\top
\end{aligned}
$$

Here, $\mathbf{Q}_i$ is the matrix representation of the quaternion $g(\underline{s}_i)$ and, again, $\mathbf{D} = \mathrm{diag}(1, -1, -1, -1)$. Finally, the desired Bingham distribution parameters are obtained from $\mathbf{C}_{i+1}^p$. This entire procedure is given in Algorithm 5.3 and visualized in Figure 5.4.

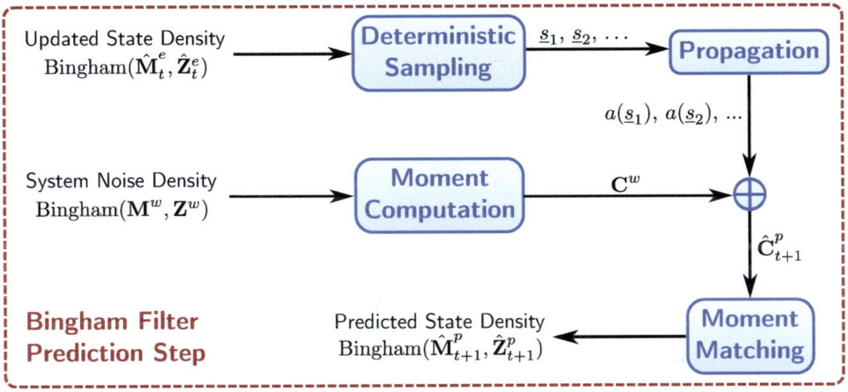

Figure 5.4.: Prediction step of Bingham filter using deterministic sampling.

If the estimate needs to be expressed in terms of a point on the hypersphere, then this point is obtained as one of the modes of the Bingham($\mathbf{M}_t^e, \mathbf{Z}_t^e$) distribution. However, care needs to be taken in the case where not only z_n, but also z_{n-1}, equal to zero because the resulting Bingham distribution exhibits circular symmetry. Thus, there is an infinite number of orientations being equally likely.

The derivation presented here serves the purpose of a better understanding of the algorithm. A more efficient way to compute $\mathbf{C}_{t+1}^p$ is computing the covariance of $\underline{w}_t$ and the sample covariance $\mathbf{C}_g$ and then making use of the fact that the covariance of a product $\underline{a} \oplus \underline{b}$ can be obtained in closed form from the covariances of $\underline{a}$ and $\underline{b}$. Furthermore, in a practical implementation the performance can be also optimized by considering antipodal symmetry and thus reducing the number of `for`-loop iterations. That is, it is sufficient to use $2n-1$ samples rather than $4n-2$.

It is also important to note that, in fact, the algorithm does not use of any properties of a Bingham distribution. In fact, the proposed procedure can be thought of as purely moment based because the proposed algorithm only uses the covariance matrices (in terms of their respective eigendecompositions) of $g(\underline{x}_t^e)$ and $\underline{w}_t$.

Algorithm 5.3 Unscented Bingham Filter Prediction

procedure $\textsc{Prediction}(\mathbf{M}_t^e,\ \mathbf{Z}_t^e,\ \mathbf{M}_t^w,\ \mathbf{Z}_t^w,\ \lambda,\ g(\cdot))$
  for $i \in \{1, ..., n\}$ **do**
    $c_i^x \leftarrow \left.\frac{\partial N(\mathbf{Z})}{\partial z_i}\right|_{\mathbf{Z}=\mathbf{Z}_t^e} \cdot N(\mathbf{Z}_t^e)^{-1};$
    $c_i^w \leftarrow \left.\frac{\partial N(\mathbf{Z})}{\partial z_i}\right|_{\mathbf{Z}=\mathbf{Z}_t^w} \cdot N(\mathbf{Z}_t^w)^{-1};$
  end for
  $\mathbf{C}_t^e \leftarrow \mathbf{M}_t^e \cdot \mathrm{diag}(c_1^x, \dots, c_n^x) \cdot (\mathbf{M}_t^e)^\top;$
  $\mathbf{C}_t^w \leftarrow \mathbf{M}_t^w \cdot \mathrm{diag}(c_1^w, \dots, c_n^w) \cdot (\mathbf{M}_t^w)^\top;$
  $(p_1, \underline{s}_1), \dots, (p_{4n-2}, \underline{s}_{4n-2}) \leftarrow \mathrm{DeterministicSampling}(\mathbf{C}_t^e, \lambda);$
  $\mathbf{C}_{t+1}^p \leftarrow \mathbf{0};$ ▷ Initilization
  $\mathbf{D} \leftarrow \mathrm{diag}(1, -1, -1, -1);$ ▷ As in (5.3)
  for $i \in \{1, ..., 4n-2\}$ **do** ▷ Second moment of $g(\underline{s}_i) \oplus \underline{w}_t$
    $\mathbf{Q} \leftarrow \mathrm{MatrixRepresentation}(g(\underline{s}_i));$ ▷ Using (5.2)
    $\mathbf{M} \leftarrow \mathbf{D} \cdot \mathbf{Q} \cdot \mathbf{D};$
    $\mathbf{C}_{t+1}^p \leftarrow \mathbf{C}_{t+1}^p + p_i \cdot \mathbf{M} \cdot \mathbf{C}_t^w \cdot \mathbf{M}^\top;$
  end for
  $(\mathbf{M}_{t+1}^p,\ \mathbf{Z}_{t+1}^p) \leftarrow \mathrm{EstimateBingham}(\mathbf{C}_{t+1}^p);$
  return $(\mathbf{M}_{t+1}^p,\ \mathbf{Z}_{t+1}^p);$
end procedure

5.6.2 Measurement Update

The Bingham parameters computed in the prediction step, will be used for the measurement update step. The moment matching procedure was computationally burdensome but it is a necessary burden because there is currently no way to perform a Bingham measurement update based on using the covariance matrix directly. To derive the measurement update, it is necessary to take a look at the inverse of uncertain quaternions which follow a Bingham$(\mathbf{M}, \mathbf{Z})$ distribution on S^3.

Lemma 5.4. *The quaternion inverse of a* Bingham$(\mathbf{M}, \mathbf{Z})$ *distributed random vector is itself a Bingham distributed random vector with dispersion parameter* $\mathbf{Z}$*, location parameter* $\overline{\mathbf{M}} := \mathbf{D} \cdot \mathbf{M}$*, and* $\mathbf{D} = \mathrm{diag}(1,\ -1,\ -1,\ -1)$.

Proof. The quaternion inverse $\underline{q}^{-1}$ of a unit quaternion $\underline{q}$ is simply given by its conjugated quaternion $\underline{q}^*$. Thus, we have

$$\begin{aligned} f((x_1, x_2, x_3, x_4)^\top; \overline{\mathbf{M}}, \mathbf{Z}) =& f(\mathbf{D} \cdot (x_1, x_2, x_3, x_4)^\top; \mathbf{M}, \mathbf{Z}) \\ =& f((x_1, -x_2, -x_3, -x_4)^\top; \mathbf{M}, \mathbf{Z}) \ . \end{aligned}$$

□

Now, the measurement equation (5.12) can be reformulated as

$$\underline{x}_t^{-1} \oplus \underline{z}_t = \underline{v}_t \ .$$

This yields

$$f(\underline{x}_t | \underline{z}_t) \propto f_{\underline{v}}(\underline{x}_t^{-1} \oplus \underline{z}_t) \cdot f(\underline{x}_t) \ .$$

The following Lemma characterizes $f_{\underline{v}}(\underline{x}_t^{-1} \oplus \underline{z}_t)$.

Lemma 5.5. *Let* $\underline{v} \sim \text{Bingham}(\mathbf{M}_v, \mathbf{Z}_v)$ *with density* $f_{\underline{v}}$. *Furthermore, let* $\underline{z} \in \mathbb{R}^4$ *with* $||\underline{z}|| = 1$. *Then* $f(\underline{x}) = f_{\underline{v}}(\underline{x}^{-1} \oplus \underline{z})$ *is the density of a* $\text{Bingham}(\underline{z} \oplus \overline{\mathbf{M}}_v, \mathbf{Z}^v)$ *distribution, where* $\underline{z}_t \oplus \overline{\mathbf{M}}_v$ *is understood as a matrix where the i-th column is the Hamilton product of* $\underline{z}$ *and the i-th column of* $\overline{\mathbf{M}}$.

Proof. First, we use (5.3) and (5.4) to rewrite $\underline{x}^{-1} \oplus \underline{z}$ as

$$\underline{x}^{-1} \oplus \underline{z} = \mathbf{D}\, (\underline{z}^{-1} \oplus \underline{x}) = \mathbf{D}\,\mathbf{D}\,\mathbf{Q}_{\underline{z}^{-1}}\,\mathbf{D}\,\underline{x} \ .$$

Next, we observe that $\det |\mathbf{D}\,\mathbf{D}\,\mathbf{Q}_{\underline{z}^{-1}}\,\mathbf{D}| = 1$ and thus, as a consequence of the transformation theorem for probability densities, $f(\underline{x})$ is an actual probability density. Finally, we make use of $\mathbf{D} = \mathbf{D}^\top$ and $\mathbf{Q}_{\underline{z}^{-1}} = \mathbf{Q}_{\underline{z}}^\top$. This yields

$$\begin{aligned} (\underline{x}^{-1} \oplus \underline{z})^\top \mathbf{M} &= (\mathbf{D}\,\mathbf{D}\,\mathbf{Q}_{\underline{z}^{-1}}\,\mathbf{D}\,\underline{x})^\top \mathbf{M} \\ &= \underline{x}^\top \mathbf{D}\,\mathbf{Q}_{\underline{z}}^\top\,\mathbf{D}\,\underbrace{\mathbf{D}\,\mathbf{M}}_{\overline{\mathbf{M}}} \\ &= \underline{x}^\top (\underline{z} \oplus \overline{\mathbf{M}}) \end{aligned}$$

and completes the proof. □

The Lemma above can be used for computing $f(\underline{z}_t|\underline{x}_t)$. This yields a Bingham distribution with parameters $\mathbf{M}_t^e$ and $\mathbf{Z}_t^e$ obtained from a eigendecomposition of

$$(\underline{z}_t \oplus \overline{\mathbf{M}}_v)\mathbf{Z}^v(\underline{z}_t \oplus \overline{\mathbf{M}}_v)^T + \mathbf{M}_t^p\mathbf{Z}_t^p(\mathbf{M}_t^p)^T$$

with $(\overline{\mathbf{M}}_v \oplus \underline{z}_t)$ denoting the matrix created by multiplying the first three lines of $\mathbf{M}^v$ with -1 and then computing the Hamilton product of $\underline{z}$ with each column.

Algorithm 5.4 Bingham Filter Update

```
1: procedure MeasurementUpdate(s_1, ..., s_m)
2:     M ← z_t ⊕ M̄_v;
3:     C ← M Z^v M^⊤ + M_t^p Z_t^p (M_t^p)^⊤;
4:     (v_1, ..., v_n, z_1, ..., z_n) ← Eigendecomposition(C);    ▷ Assuming
   z_i ≤ z_{i+1};
5:     M_t^e ← (v_1, ..., v_n);
6:     Z_t^e ← diag(z_1 − z_n, ... z_{n−1} − z_n, 0);
7:     return (M_t^e, Z_t^e);
8: end procedure
```

5.7 Evaluation

The methods discussed in this chapter are evaluated in three ways. First, we evaluate the computation of the normalization constant and the resulting parameter estimation procedure because these are the driving factors behind the computational cost of the Bingham filter. Second, we evaluate the quality of propagation using the proposed deterministic sampling scheme. Finally, we compare the entire resulting filter with the UKF and a particle filter.

5.7.1 Normalization Constant and Parameter Estimation

Our evaluation of the computation of the normalization constant was carried out for the Bingham distribution on S^3 because this case is of particular interest in our work. We evaluated the normalization constant $N(\mathbf{Z})$

for different values of $\mathbf{Z}$. This evaluation was subdivided into three cases involving a different number of entries in $\mathbf{Z}$ significantly different from zero. These cases were $\mathbf{Z} = \text{diag}(-a, -a, -a, 0)$, $\mathbf{Z} = \text{diag}(-a, -a, 0.1, 0)$, and $\mathbf{Z} = \text{diag}(-a, 0.1, 0.1, 0)$. This distinction accounts for the fact that computational performance of certain methods is dependent on the structure of $\mathbf{Z}$. The use of 0.1 is due to the fact that the original implementation of some comparison methods are incapable of handling multiple zero entries. Ground truth was obtained using numerical integration. In particular, Matlab's `integral3` function was used to compute the triple integral.

For comparing the computation of the normalization constant, we used the series expansion (which was also used to generate the precomputed lookup tables in `libbingham`) from [34], the approach based on holonomic gradient descent [95], and the algorithm proposed by Koev [66]. Results of this comparison in terms of accuracy and computation time are shown in Figure 5.5. The method by Koev was called using `mhg(100,2,0.5,2,Z)`.

For evaluating parameter estimation, the same matrices were used as in the evaluation of the normalization constant. Three different methods were compared. The first approach used a Gauss-Newton method implemented in Matlab. The second approach used a Gauss-Newton method implemented in C. Finally, the third approach was based on the `fsolve` function in Matlab, which uses a trust-region algorithm. Deterministic sampling was used to generate samples corresponding to the desired ground truth parameters. That is, the second moment of the generated sample set corresponds to the desired second moments of the Bingham distribution. The saddlepoint-approximation-based method was used for computing the Bingham normalization constant and its derivatives in all three methods. The results are shown in Figure 5.6. Even though the approach based on `fsolve` is much more precise, use of the computationally less burdensome Gauss-Newton approximation seems sufficient for most applications involving high noise levels as the relative error introduced by the choice of the optimization procedure is below 1%, and thus, negligible compared to the uncertainty of the estimate.

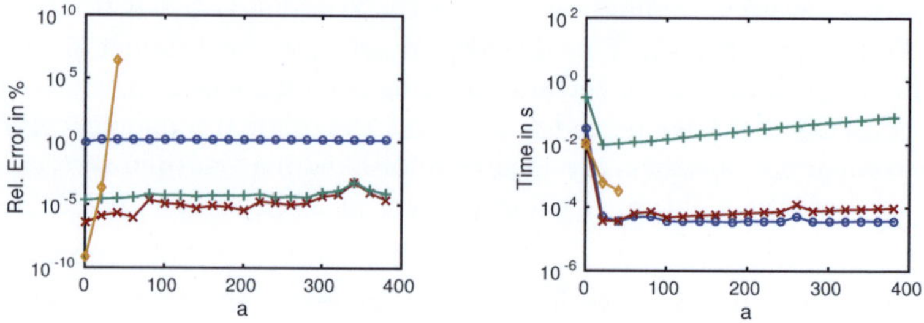

(a) $\mathbf{Z} = -\mathrm{diag}(a, a, a, 0)$.

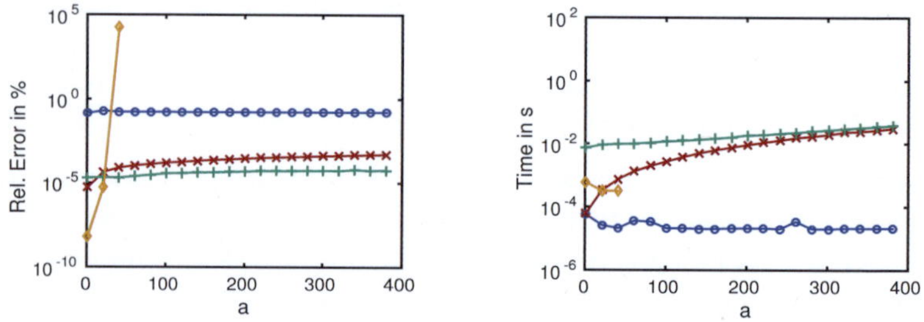

(b) $\mathbf{Z} = -\mathrm{diag}(a, a, 0.1, 0)$.

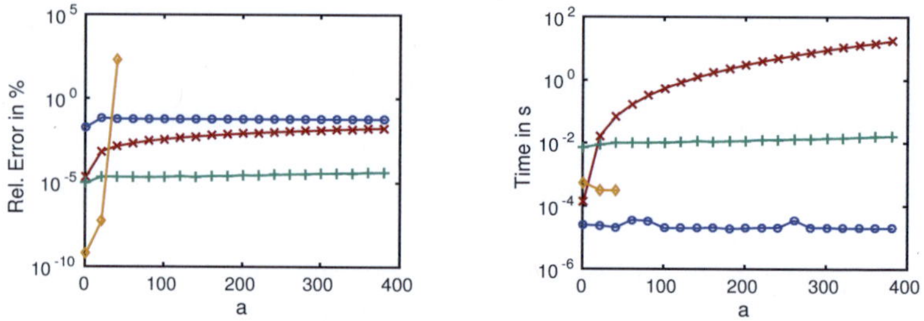

(c) $\mathbf{Z} = -\mathrm{diag}(a, 0.1, 0.1, 0)$.

Figure 5.5.: Accuracy (left column) and computation time (right column) for computing the Bingham normalizing constant $N(\mathbf{Z})$ using the saddlepoint approximations based approach (blue), the algorithm proposed by Koev (orange), the series expansion used by Glover (red), and holonomic gradient descent (green).

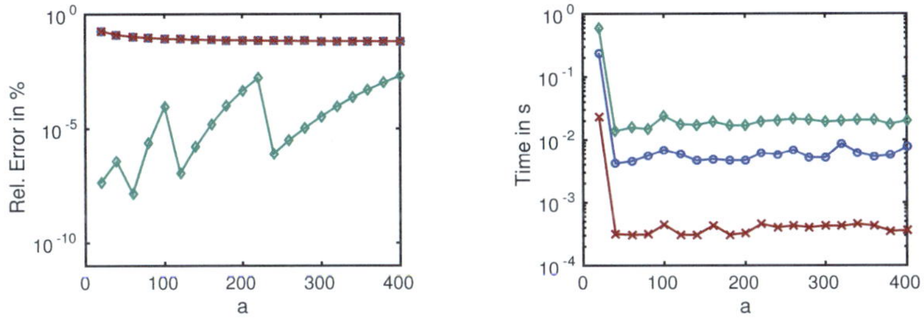

(a) $\mathbf{Z} = -\mathrm{diag}(a, a, a, 0)$.

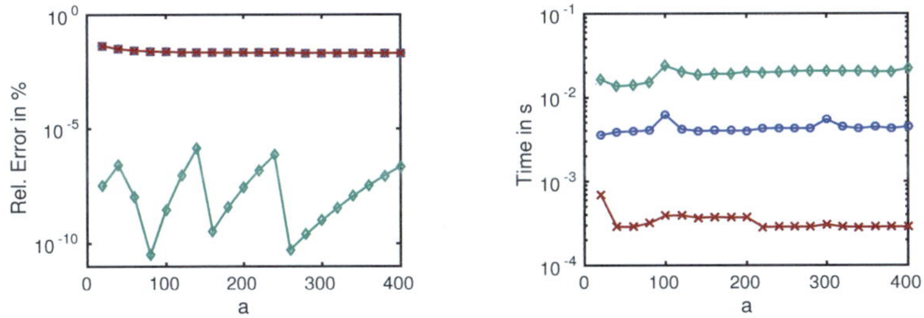

(b) $\mathbf{Z} = -\mathrm{diag}(a, a, 0.1, 0)$.

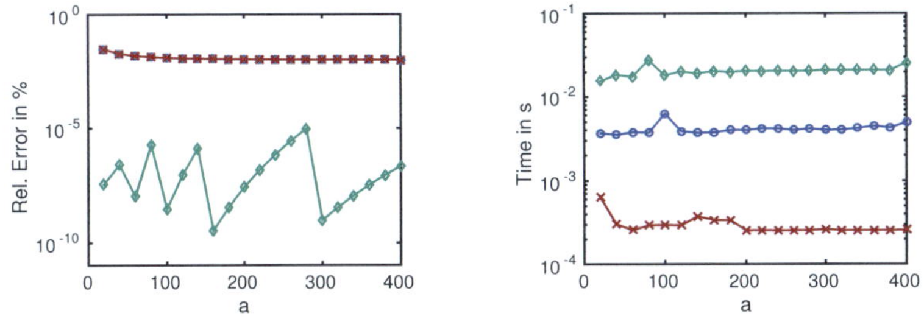

(c) $\mathbf{Z} = -\mathrm{diag}(a, 0.1, 0.1, 0)$.

Figure 5.6.: Accuracy (left column) and computation time (right column) for computing the Bingham MLE using a Gauss-Newton algorithm implemented in C (red) and MATLAB (blue), and an alternative based on `fsolve` (green). The accuracy of the Gauss-Newton method is equal for both implementations, and thus, the curves are indistinguishable.

5.7.2 Propagation

For evaluating the propagation based on the deterministic sampling scheme proposed in this chapter, it is of particular interest to use a system function that does not preserve the Bingham distribution. Thus, quaternion spherical linear interpolation (SLERP) [109] was used for this comparison. It is defined by

$$g(\underline{x}) = \underline{x}^b \oplus \underline{q}^{(1-b)} \ ,$$

where $\underline{x}^b$ denotes quaternion exponentiation and $a \in (0, 1)$. This mapping is usually used to interpolate orientations. Here, it was implemented in a slightly modified way respecting antipodal symmetry. The goal of the evaluation procedure is the computation of $\hat{\mathbf{C}} = \mathbb{E}(g(\underline{x}) \cdot g(\underline{x})^\top)$ where $\underline{x}$ is a Bingham$(\mathbf{I}, \mathbf{Z})$ distributed random vector (using different values in $\mathbf{Z}$). In the system function, $\underline{q}$ was chosen as $(0.5, 0.5, 0.5, 0.5)^\top$. Ground truth was generated by using random sampling with 10^6 samples. The eigenvector of $\hat{\mathbf{C}}$ corresponding to the largest eigenvalue was used as the estimate of the mean. The results were compared using the angle of the Rodrigues' rotation formula. For two orientations $\underline{a}$ to $\underline{b}$, this angle is given by

$$r(\underline{a}, \underline{b}) := 2 \cdot \min(\operatorname{acos}(\underline{a}^\top \underline{b}), \pi - \operatorname{acos}(\underline{a}^\top \underline{b})) \ .$$

It can be thought of as the angle between these two orientations. The results that show the angular mean error are visualized in Figure 5.7. For these scenarios, the error usually stays below 0.1°. However, it may become even larger when $\mathbf{Z}$ approaches the zero matrix.

5.7.3 Filter

Finally, the entire proposed filter was compared against a modified UKF and a particle filter. Once again, SLERP was used as the system function, where $\underline{q}$ was chosen as in the propagation evaluation and $a = 0.1$. This was corrupted by some quaternion noise which resulted in the model

$$\underline{x}_{t+1} = g(\underline{x}_t) \oplus \frac{\underline{w}_t}{||\underline{w}_t||} \ .$$

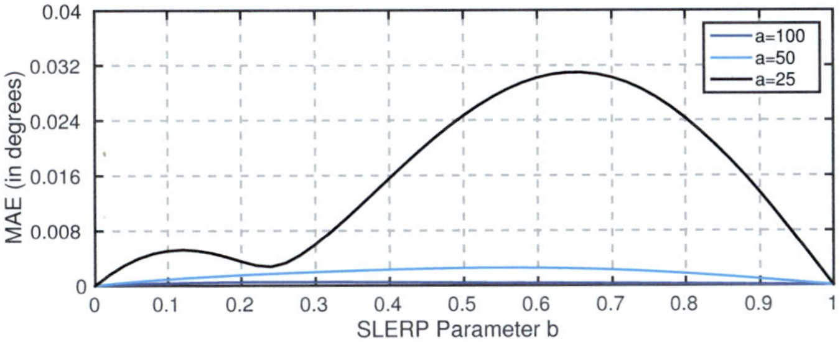

(a) Deterministic sampling parameter $\lambda = 0.1$.

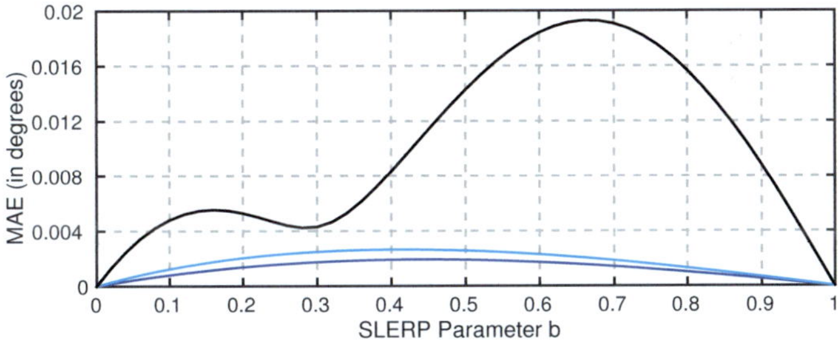

(b) Deterministic sampling parameter $\lambda = 0.5$.

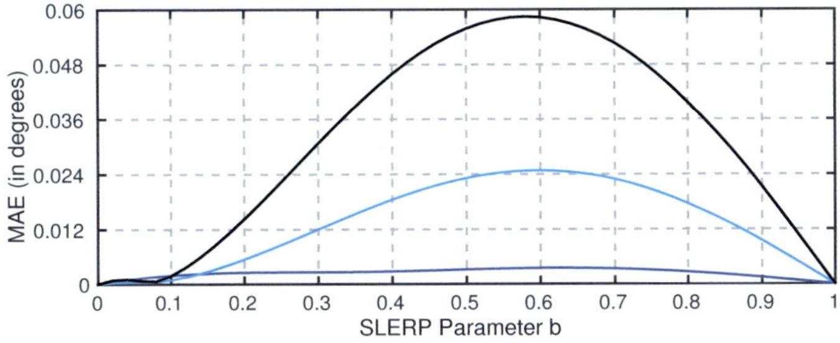

(c) Deterministic sampling parameter $\lambda = 0.9$.

Figure 5.7.: Mean angular error (MAE) in degrees when propagating a Bingham uncertainty through the SLERP system function with parameter a. The considered Bingham distribution is given by $\mathbf{M} = \mathbf{I}$ and $\mathbf{Z} = -\text{diag}(a, a, a, 0)$ with $a = 25$ (black), $a = 50$ (blue), $a = 100$ (dark blue).

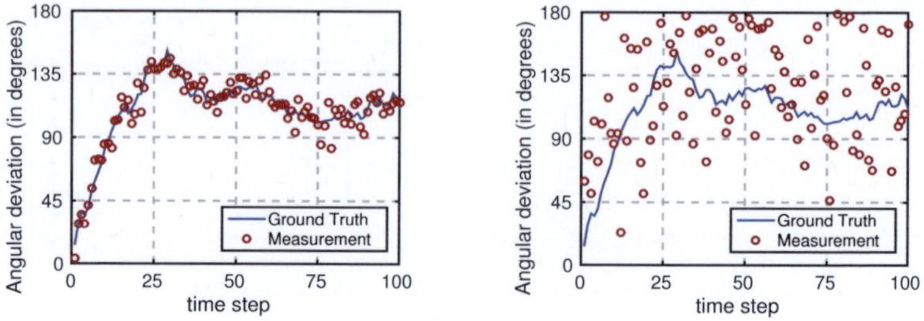

(a) A typical run and the corresponding measurements.

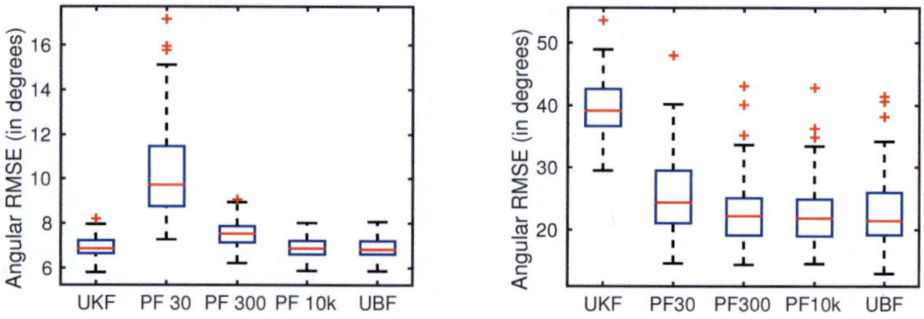

(b) Angular RMSE for different types of filters.

Figure 5.8.: Evaluation of the filter for scenarios involving low measurement noise (left column) and high measurement noise (right column).

The measurement model considers noisy direct measurements, i.e., it is given by

$$\underline{z}_t = \underline{x}_t \oplus \frac{\underline{v}_t}{||\underline{v}_t||} \ .$$

To avoid giving the Bingham filter an advantage, the Gaussian distribution was used for generating system and measurement noise[1], and thus $\underline{w}_t$ was replaced by $\underline{w}_t \cdot ||\underline{w}_t||^{-1}$ and $\underline{v}_t$ was replaced by $\underline{v}_t \cdot ||\underline{v}_t||^{-1}$. These replacements are necessary in order to ensure the noise to be defined on the manifold. The modification of the UKF was twofold. First, an intelligent repositioning of measurements was introduced ensuring them

[1] Simulations using the Bingham distribution for generating the noise gave similar results.

to be on the same side of the hypersphere as the current estimate. That is, $\underline{z}_t$ was replaced by $-\underline{z}_t$ if $||\underline{z}_t - \underline{x}_t^e|| > ||-\underline{z}_t - \underline{x}_t^e||$. Second, after the measurement update step, a projection was performed (dividing the obtained estimate by its norm) ensuring the finally resulting quaternion to have unit length. An approximation was used to compute the likelihood of the particle filter by assuming $p(\underline{z}_t|\underline{x}_t) \approx f_v((\underline{x}_t^p)^{-1} \oplus \underline{z}_t)$, where f_v is the Gaussian pdf of $\underline{v}_t$ rather than the pdf of $\underline{v}_t \cdot ||\underline{v}_t||^{-1}$.

The evaluation consisted of 100 runs with 100 time steps per run. Parameters for generating the initial position and the noise had mean $\underline{\mu}_0 = \underline{\mu}_w = \underline{\mu}_v = (1, 0, 0, 0)$, which corresponds to the vector representation of the quaternion 1, i.e., zero mean in the orientation sense. The covariances corresponded to an expected angular deviation (computed using the angular measure from the propagation simulation) of 18.2° for the initial position and 5.8° for the system noise. This was achieved by choosing $\mathbf{C}_0 = 0.01 \cdot \mathbf{I}$ and $\mathbf{C}_w = 0.001 \cdot \mathbf{I}$. Two scenarios were considered for the measurement noise. For the high noise scenario, $\mathbf{C}_v = 0.3 \cdot \mathbf{I}$ was chosen (corresponding to an expected angular deviation of 85.8°) and, analogously, $\mathbf{C}_v = 0.003 \cdot \mathbf{I}$ was used for the low noise scenario (corresponding to an expected angular deviation of 10.0°).

The initial estimates for the filters were given by a Gaussian with parameters $\underline{\mu} = (0, 0, 0, 1)^\top$ and $\mathbf{C}_v = \mathbf{I}$. Corresponding Bingham distribution parameters for the proposed filter were obtained by random sampling from the Gaussian distributions involved. Renormalization to unit length and a subsequent matching of Bingham distribution parameters. Three variants of the particle filter were used, which differed in the number of particles (30, 300, and 10.000). The entire results of the evaluation are visualized in Figure 5.8. As expected, in the case of small noise, the filters making a wrong Gaussian assumption perform similarly to the Bingham filter. However, as the noise becomes larger, the Bingham filter significantly outperforms the UKF and the particle filter with 30 particles. Similar results are achieved when increasing the number of particles to 300. A further increase seems to have no significant effect on the quality of the results.

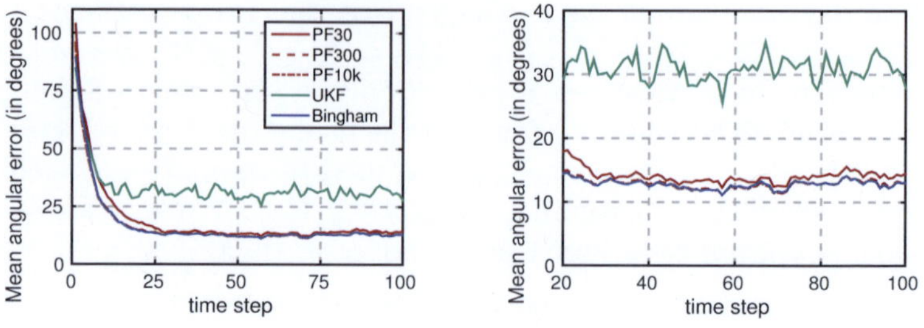

(a) Entire run. **(b)** After filter initialization phase.

Figure 5.9.: Mean angular error of the high noise scenario.

5.8 Summary and Discussion

In this chapter a novel method for orientation estimation from noisy measurements is proposed. Its main theoretical contribution is a deterministic sampling scheme for antipodally symmetric distributions defined on the hypersphere. The proposed method is based on moment matching, and thus, can be thought of as a hyperspherical equivalent of the UKF. This deterministic sampling scheme is used as the basis for a novel stochastic filter based on the Bingham distribution, which is an antipodally symmetric distribution defined on the hypersphere. In case of the 4d unit sphere, this distribution can be used to represent uncertain orientations based on unit quaternions. The resulting filtering technique is also capable of a closed-form measurement update in case of noisy direct measurements. The involved moment matching procedures require multiple evaluations of the Bingham normalization constant, thus we make use of saddlepoint approximations to speed up this procedure. We also show that the resulting optimization problem is convex. In the simulations, it is shown that the proposed filtering scheme outperforms approaches that implicitly assume linearity.

CHAPTER 6

Estimation of Planar Rigid-Body Motions

Contents

6.1 Introduction

In the preceding chapters, we considered either uncertain quantities defined on a linear space or uncertain quantities defined on the hypersphere.

This chapter discusses simultaneous consideration of linear and directional quantities. Particularly, we focus on the estimation of position and orientation in the plane, which can also be understood as estimation of planar rigid-body motions.

Estimating these quantities is of particular interest for several technical applications in fields such as robotic perception or mixed and augmented reality. Applications include (but are not limited to) pose estimation, sensor calibration, and handling of weak features in vision systems. Use of combined models for simultaneous consideration of position and orientation is necessary when these two quantities cannot be assumed to be independent from each other. Once again, as long as all arising uncertainties are small, the entire (partially nonlinear state space) can be approximated by a linear space making use of local linearity of the underlying domain. However, this approach might yield suboptimal or even entirely wrong results whenever high uncertainties are involved. Thus, we propose an approach that systematically accounts for dependencies between position and orientation.

The first contribution of this chapter is a novel probability distribution belonging to the exponential family of distributions, which naturally arises when conditioning a sub-vector of a zero-mean Gaussian random vector to unit length. The novel distribution can be used for representation of rigid-body motions in the plane. Additionally, we derive a deterministic sampling scheme that differs from both approaches presented in previous chapters while making use of them in special cases. The deterministic sampling approach in Chapter 4 was based on approximating the shape of a continuous density, and thus, was distribution dependent, whereas the approach in the preceding chapter was purely based on a moment matching procedure, and thus, can be thought of as independent of the underlying distribution. The work presented in this chapter is hybrid in the sense that it considers the actual distribution structure but may involve moment matching or arbitrary other discrete approximation procedures. All of this is used to propose a filter that truly considers the structure of the manifold of rigid-body motions.

The remainder of this chapter, which is based on and extends our work in [138], is structured as follows. In the next section, we discuss some related approaches which involve the use of directional statistics or dual quaternions for estimation of rigid-body motions. Sec. 6.3 contains an introduction to dual quaternions and a derivation of a multiplicative subgroup for representing planar rigid-body motions. The new probability distribution is introduced in Sec. 6.4 by discussing marginal and conditional distributions, describing a parameter estimation procedure, and discussing its application to representation of uncertain planar rigid-body motions. A deterministic sampling scheme is proposed in Sec. 6.5. It makes use of the structure of the proposed distribution, and we show how it can be derived from sampling schemes of the Bingham and the Gaussian distribution. These results are brought together for the derivation of a novel filter in Sec. 6.6, which is evaluated in Sec. 6.7. The proposed methodology is summarized in Sec. 6.8.

6.2 Related Approaches

The use of dual quaternions for representing the group of rigid-body motions has been considered in many works on computational perception and kinematics. An introduction from the viewpoint of matrix algebra is given in [117]. An extrinsic calibration procedure was proposed in [11] where the relative pose of two sensors is estimated. Skinning with dual quaternions is discussed in [61] and [62]. A linearization based approach of dynamic state estimation using dual quaternions has been taken in [38], which uses an EKF for state estimation. An optimization based estimation approach has been proposed in [121]. Using dual quaternions in Gaussian process regression has been proposed in [79].

Unfortunately, there is only a very limited amount of work discussing probability distributions simultaneously considering periodic and non-periodic quantities. A matrix based equivalent to the distribution proposed in this work was originally presented in [75]. Another approach is wrapping of certain dimensions of a normal distributed random vector. This can be thought as a generalization of both, the Gaussian distribution and a multivariate wrapped normal distribution (the special case of a bivariate wrapped normal is addressed in [102]). The resulting partially wrapped normal distribution was discussed in [147]. This is a generalization of a first work considering a normal distributed random vector with one wrapped dimension [51].

So far, there has been only one approach combining dual quaternion representation with directional statistics in [27], [80], [28]. It is based on making use of a projected Gaussian distribution in the sense that it uses a Gaussian defined on the tangent space of a point on the manifold of rigid-body motions and then projects the probability mass from this tangent space to other points on the manifold. It is important to note, that this terminology is somewhat ambiguous, because the terminology "Projected Gaussian" sometimes refers to the distribution of $\underline{v} \cdot ||\underline{v}||^{-1}$, where $\underline{v}$ is a Gaussian distributed random vector.

6.3 Dual Quaternions for Representing Rigid-Body Motions in the Plane

Dual quaternions generalize both quaternions and dual numbers. They are of our interest because of their capability of representing rigid-body motions systematically. Thus, we will give a brief introduction to dual quaternions and then discuss their application to representing rigid-body motions and the special case considered here, planar rigid-body motions.

6.3.1 Fundamental Properties

As quaternions have been introduced in the previous chapter, we only introduce the concept of dual numbers before laying out how both concepts are combined in order to obtain dual quaternion numbers.

Dual Numbers

The concept of dual numbers was developed by Clifford in [20]. A dual number can be seen as a modification of complex numbers replacing the complex unit i by the dual unit ε, which is characterized by the nilpotency property, that is $\varepsilon^2 = 0$. Thus, a dual number can be written as

$$a_1 + a_2 \, \varepsilon \; .$$

The sum of two dual numbers is simply given as the componentwise sum. Multiplication follows from the nilpotency property as

$$(a_1 + a_2\varepsilon) \cdot (b_1 + b_2\varepsilon) = a_1 b_1 + (a_1 b_2 + a_2 b_1)\varepsilon \ .$$

A matrix representation of a dual number $a_1 + a_2\varepsilon$ is given as

$$\begin{pmatrix} a_1 & 0 \\ a_2 & a_1 \end{pmatrix} . \tag{6.1}$$

With this representation, matrix multiplication can be used to multiply two dual numbers. Furthermore, we can define a dual conjugation by changing the sign before the dual part. However, it is easily seen that a dual number does not necessarily have a multiplicative inverse. Whenever the real part is zero, a dual number is not invertible.

Dual Quaternions

A dual quaternion can be thought of as a dual number where both entries are replaced by quaternions or equivalently as a quaternion where all entries are replaced by dual numbers. That is, a dual quaternion is given by an 8-tuple of real values

$$(a_1 + a_2 i + a_3 j + a_4 k) + \varepsilon(a_5 + a_6 i + a_7 j + a_8 k) \ . \tag{6.2}$$

Moreover, the dual unit ε commutes with i, j, k, e.g., it holds $i\varepsilon = \varepsilon i$. One could also define dual quaternions in a different way, where the relationship between the quaternion units and the dual unit anticommutes, i.e. $\varepsilon i = -i\varepsilon$. This variant is often used in the special case of defining dual complex numbers for applying them to the representation of planar rigid-body motions. However, in this work, dual quaternions are chosen rather than dual complex numbers for better consistency with future generalizations to rigid-body motions in $\mathbb{R}^3$.

Multiplication of dual quaternions is obtained directly from multiplication of dual numbers and multiplication of quaternions. Let $\underline{a}, \underline{b} \in \mathbb{R}^8$ denote dual quaternions (expressed as vectors). These vectors can be split up into sub-vectors representing quaternions $\underline{a}_q, \underline{a}_d, \underline{b}_q, \underline{b}_d \in \mathbb{R}^4$. Then, the dual quaternion product (denoted by $\boxplus$) can be expressed as

$$\begin{aligned}
\underline{a} \boxplus \underline{b} &= \begin{pmatrix} \underline{a}_q \\ \underline{a}_d \end{pmatrix} \boxplus \begin{pmatrix} \underline{b}_q \\ \underline{b}_d \end{pmatrix} \\
&= \begin{pmatrix} \underline{a}_q \oplus \underline{b}_q \\ \underline{a}_q \oplus \underline{b}_d + \underline{a}_d \oplus \underline{b}_q \end{pmatrix} \\
&= \begin{pmatrix} a_1 b_1 - a_2 b_2 - a_3 b_3 - a_4 b_4 \\ a_1 b_2 + a_2 b_1 + a_3 b_4 - a_4 b_3 \\ a_1 b_3 - a_2 b_4 + a_3 b_1 + a_4 b_2 \\ a_1 b_4 + a_2 b_3 - a_3 b_2 + a_4 b_1 \\ a_1 b_5 - a_2 b_6 - a_3 b_7 - a_4 b_8 + a_1 b_1 - a_2 b_2 - a_3 b_3 - a_4 b_4 \\ a_1 b_6 + a_2 b_5 + a_3 b_8 - a_4 b_7 + a_1 b_2 + a_2 b_1 + a_3 b_4 - a_4 b_3 \\ a_1 b_7 - a_2 b_8 + a_3 b_5 + a_4 b_6 + a_1 b_3 - a_2 b_4 + a_3 b_1 + a_4 b_2 \\ a_1 b_8 + a_2 b_7 - a_3 b_6 + a_4 b_5 + a_1 b_4 + a_2 b_3 - a_3 b_2 + a_4 b_1 \end{pmatrix} .
\end{aligned}$$

There are several ways to define the conjugation of dual quaternions. First, it is possible to conjugate the units i, j, k resulting in

$$\begin{aligned}
&((a_1 + a_2 i + a_3 j + a_4 k) + \varepsilon(a_5 + a_6 i + a_7 j + a_8 k))^{\ddagger} \\
&:= (a_1 + a_2 i + a_3 j + a_4 k)^* + \varepsilon(a_5 + a_6 i + a_7 j + a_8 k)^* \\
&= (a_1 - a_2 i - a_3 j - a_4 k) + \varepsilon(a_5 - a_6 i - a_7 j - a_8 k) ,
\end{aligned}$$

where * denotes the quaternion conjugation. As will be seen, this definition is of particular importance when using dual quaternions in context of representing rigid-body motions. Second, dual conjugation can also be applied changing the sign in front of the dual unit

$$\begin{aligned}
&((a_1 + a_2 i + a_3 j + a_4 k) + \varepsilon(a_5 + a_6 i + a_7 j + a_8 k))^{\dagger} \\
&:= ((a_1 + a_2 i + a_3 j + a_4 k) - \varepsilon(a_5 + a_6 i + a_7 j + a_8 k)) .
\end{aligned}$$

Finally, both conjugations can be applied simultaneously

$$\begin{aligned}
&((a_1 + a_2 i + a_3 j + a_4 k) + \varepsilon(a_5 + a_6 i + a_7 j + a_8 k))^{\ddagger\dagger} \\
&:= ((a_1 - a_2 i - a_3 j - a_4 k) - \varepsilon(a_5 - a_6 i - a_7 j - a_8 k)) .
\end{aligned}$$

The inverse of a dual quaternion $a + b\varepsilon$ (where $a, b \in \mathbb{R}^4$) is obtained by using the inverse of quaternions as

$$a^{-1}(1 - \varepsilon b a^{-1}) .$$

As already mentioned in the above discussion of dual numbers, it can now easily be seen that the inverse does not necessarily exist. A necessary (and sufficient) condition is $a \neq 0$. The norm of a dual quaternion $\underline{a}$ is usually defined as $\underline{a} \boxplus \underline{a}^{\ddagger}$. If this norm is equal to 1, then $\underline{a}$ is called a unit dual quaternion. However, contrary to the quaternion case, it is important to note that this norm is not necessarily a real number, as is shown in the following example.

Example 6.1. *Consider the dual quaternion* $\underline{a} = (1,0,0,0,1,0,0,0)^\top$. *Then the norm is obtained as*

$$\underline{a} \boxplus \underline{a}^{\ddagger} = (1,0,0,0,2,0,0,0)^\top \ .$$

Thus, with this norm mapping, the vectorspace of dual quaternions cannot be considered a Banachspace. However, due to its relation to norms on quaternions and complex numbers this mapping is usually still called a norm. It will play an important role when dual complex numbers are applied to representation of uncertain rigid-body motions in the plane.

Using the typical matrix representation of a quaternion (5.2) and (6.1), we can obtain a matrix representation of the dual complex number given in (6.2) as

$$\mathbf{Q}_a = \begin{pmatrix} \mathbf{Q}_{a,q} & \mathbf{0} \\ \mathbf{Q}_{a,d} & \mathbf{Q}_{a,q} \end{pmatrix} = \left(\begin{array}{cccc|cccc} a_1 & a_2 & a_3 & a_4 & 0 & 0 & 0 & 0 \\ -a_2 & a_1 & -a_4 & a_3 & 0 & 0 & 0 & 0 \\ -a_3 & a_4 & a_1 & -a_2 & 0 & 0 & 0 & 0 \\ -a_4 & -a_3 & a_2 & a_1 & 0 & 0 & 0 & 0 \\ \hline a_5 & a_6 & a_7 & a_8 & a_1 & a_2 & a_3 & a_4 \\ -a_6 & a_5 & -a_8 & a_7 & -a_2 & a_1 & -a_4 & a_3 \\ -a_7 & a_8 & a_5 & -a_6 & -a_3 & a_4 & a_1 & -a_2 \\ -a_8 & -a_7 & a_6 & a_5 & -a_4 & -a_3 & a_2 & a_1 \end{array}\right) \tag{6.3}$$

Where $\mathbf{Q}_{a,q}$ is the matrix representation of the quaternion $(a_1, a_2, a_3, a_4)^\top$ and $\mathbf{Q}_{a,d}$ is the matrix representation of the quaternion $(a_5, a_6, a_7, a_8)^\top$.

6.3.2 Representing Rigid-Body Motions

The application of unit dual quaternions to representing rigid-body motions is straightforward. A rotation with angle θ around the (normalized) axes $(v_x, v_y, v_z)^\top$ is represented in the same way as in orientation representation with pure quaternions. That is, it is given by the

$$\cos\left(\frac{\theta}{2}\right) + (v_x\, i + v_y\, j + v_z\, k) \sin\left(\frac{\theta}{2}\right) .$$

Translations (t_x, t_y, t_z) can be represented using

$$1 + \frac{\varepsilon}{2}(t_x\, i + t_y\, j + t_z\, k) .$$

The entire rigid-body motion is represented as a product of these two types of dual quaternions. Applying a rigid-body motion represented by $\underline{v} \in \mathbb{R}^8$ to a vector $\underline{w} = (w_x, w_y, w_z)$ starts of by representing this vector as a dual quaternion

$$1 + \varepsilon(w_x\, i + w_y\, j + w_z\, k)$$

and then computing $\underline{v} \boxplus \hat{\underline{w}} \boxplus \underline{v}^\ddagger$, where $\hat{\underline{w}} \in \mathbb{R}^8$ is the vector representation of the dual-quaternion representing $\underline{w}$. The result is then given in the imaginary entries of the dual part. From this, it is also seen why the dual quaternions $\underline{v}$ and $-\underline{v}$ represent the same rigid-body motion. This change of signs cancels out during computation. Besides that, the representation is unique. Thus, unit dual quaternions are a double cover of $SE(3)$.

Restricting the representation to representing purely planar body motions yields some simplifications. We use the dual quaternion framework and consider the x–y–plane. In this setting, a rotation (around the origin) with angle θ can be thought of as a rotation around $(0, 0, 1)^\top$. This is represented by the dual quaternion

$$\cos\left(\frac{\theta}{2}\right) + \sin\left(\frac{\theta}{2}\right) k .$$

Similarly, a translation $(t_x, t_y)^\top$ in the plane is represented by the dual quaternion representing the translation $(t_x, t_y, 0)^\top$. That is,

$$1 + \frac{\varepsilon}{2}(t_x\, i + t_y\, j) .$$

We can combine rotation and translation, represented by dual quaternions r and t respectively as

$$\begin{aligned} t \cdot r = &\left(1 + \frac{\varepsilon}{2}(t_x\, i + t_y\, j)\right) \cdot \left(\cos\left(\frac{\theta}{2}\right) + \sin\left(\frac{\theta}{2}\right) k\right) \\ = &\cos\left(\frac{\theta}{2}\right) + \sin\left(\frac{\theta}{2}\right) k \\ &+ \frac{\varepsilon}{2}\left[\left(\cos\left(\frac{\theta}{2}\right) t_x + \sin\left(\frac{\theta}{2}\right) t_y\right) i \right. \\ &\qquad \left. + \left(-\sin\left(\frac{\theta}{2}\right) t_x + \cos\left(\frac{\theta}{2}\right) t_y\right) j\right], \end{aligned}$$

which assumes the rotation to be carried out first. Furthermore, every planar rigid-body motion can be represented as a rotation with a subsequent translation. Thus, all dual quaternions used for representing planar rigid-body motions only have four non-zero entries. Furthermore, there is no constraint on the non-zero entries in the dual part, whereas the non-dual part needs to be a unit quaternion.

Consequently, our representation of the dual quaternion

$$(a_1 + a_2\, k) + \varepsilon(a_3\, i + a_4\, j)$$

can be simplified to using a vector in $S^1 \times \mathbb{R}^2 \subset \mathbb{R}^4$. The matrix representation (6.3) can also be simplified to

$$\mathbf{Q}_a = \begin{pmatrix} a_1 & a_2 & 0 & 0 \\ -a_2 & a_1 & 0 & 0 \\ -a_3 & a_4 & a_1 & -a_2 \\ -a_4 & -a_3 & a_2 & a_1 \end{pmatrix}$$

This matrix representation respects the multiplicative subgroup of unit dual quaternions restricted to planar motions. It corresponds to the usual matrix representation of dual complex numbers. And thus, unit dual complex numbers are a representation of this subgroup. Their application to representing planar rigid-body motions is discussed in [92].

For our reduced vector notation, we will continue to use $\boxplus$ to denote the dual quaternion product. That is,

$$\begin{pmatrix} a_1 \\ a_2 \\ a_3 \\ a_4 \end{pmatrix} \boxplus \begin{pmatrix} b_1 \\ b_2 \\ b_3 \\ b_4 \end{pmatrix} = \begin{pmatrix} a_1 b_1 - a_2 b_2 \\ a_1 b_2 + a_2 b_1 \\ a_1 b_3 - a_2 b_4 + a_3 b_1 + a_4 b_2 \\ a_1 b_4 + a_2 b_3 - a_3 b_2 + a_4 b_1 \end{pmatrix} .$$

Similar to the previous chapter, we can rewrite some operations as matrix vector multiplications. The multiplication can be rewritten as

$$\underline{a} \boxplus \underline{b} = \mathbf{D}\,\mathbf{Q}_a\,\mathbf{D}\,\underline{b} \,, \tag{6.4}$$

where once again $\mathbf{D} = \operatorname{diag}(1, -1, -1, -1)$. The same can be done for inversion, because this comes down to a conjugation in the case considered here. Thus, the inverse of $\underline{a}$ is given as

$$\underline{a}^{-1} = \underline{a}^{\ddagger} = \mathbf{D}\,\underline{a} \,. \tag{6.5}$$

6.4 A New Probability Distribution

In order to be capable of representing uncertain planar rigid-body motions, we propose a novel probability distribution that represents uncertainty over the multiplicative subgroup of unit dual quaternions as derived above. That is, we define a probability distribution on $\Omega = S^1 \times \mathbb{R}^2$. When interpreting the circular part as a unit vector in $\mathbb{R}^2$, we have $\Omega \subset \mathbb{R}^4$.

The basic idea of the proposed distribution defined on Ω is similar to the Bingham distribution. That is, it is basically a conditioned multivariate Gaussian. In case of the Bingham distribution, the entire random vector is conditioned to unit length. The proposed distribution is slightly different as only a sub-vector is conditioned to unit length. In the considered case, this condition is imposed on the sub-vector consisting of the first two entries of $\underline{x}$. Thus, the probability density of the newly proposed distribution is given as

$$f(\underline{x}) = K(\mathbf{P})^{-1} \exp\left(\underline{x}^\top \mathbf{P}\,\underline{x}\right) \tag{6.6}$$

with $\underline{x} \in \Omega$, a parameter matrix $\mathbf{P}$, and a normalization constant $K(\mathbf{P})$. So far, it is seen that the newly proposed density is, similarly to the Bingham case, antipodally-symmetric. Thus, $\mathbb{E}(\underline{x}) = \underline{0}$ holds independently of the choice of $\mathbf{P}$.

We can rewrite $\underline{x}$ as $(\underline{x}_s^\top, \underline{x}_t^\top)^\top$ with $\underline{x}_s \in S^1$, $\underline{x}_t \in \mathbb{R}^2$. This notation will be useful in the following discussion of the proposed distribution. A direct numerical integration for computing the normalization constant $N(\mathbf{P})$ is as complicated as in the Bingham case, because it requires the computation of the triple integral

$$K(\mathbf{P}) = \int_{S^1} \int_{\mathbb{R}^2} \exp\left((\underline{x}_s^\top, \underline{x}_t^\top)\mathbf{P}(\underline{x}_s^\top, \underline{x}_t^\top)^\top\right) \, \mathrm{d}\underline{x}_t \, \mathrm{d}\underline{x}_s \ .$$

However, later it will be shown how this computation can be simplified to the special case of computing a Bingham normalization constant. Another problem of the current formulation is the fact, that it is not obvious which conditions need to be satisfied by $\mathbf{P}$ in order to ensure that the proposed density is a well defined probability distribution. In what follows, $\mathbf{P}$ is assumed to be symmetric. Obviously, it is sufficient that $\mathbf{P}$ is negative-definite. But it is not clear whether negative-definiteness of $\mathbf{P}$ is also a necessary condition.

The first step for better understanding the proposed distribution is rewriting the parameter matrix $\mathbf{P}$ as

$$\mathbf{P} = \begin{pmatrix} \mathbf{P}_1 & \mathbf{P}_2^\top \\ \mathbf{P}_2 & \mathbf{P}_3 \end{pmatrix} \ . \tag{6.7}$$

With this notation, we can obtain a more convenient formulation of the considered density (6.6).

Proposition 6.2. *The density* (6.6) *can be formulated as*

$$f(\underline{x}_s, \underline{x}_t) = \frac{1}{K(\mathbf{P})} \exp\left(\underline{x}_s^\top \, \mathbf{S} \, \underline{x}_s + (\underline{x}_t - \mathbf{A}\underline{x}_s)^\top \mathbf{P}_3 (\underline{x}_t - \mathbf{A}\underline{x}_s)\right) \ , \tag{6.8}$$

where $\mathbf{P}_i$ *are as in* (6.7), $\mathbf{S} = \mathbf{P}_1 - \mathbf{P}_2^\top \mathbf{P}_3^{-1} \mathbf{P}_2$ *is the Schur complement of* $\mathbf{P}_3$ *in* $\mathbf{P}$, *and* $\mathbf{A} = -\mathbf{P}_3^{-1}\mathbf{P}_2$.

Proof. We make use of the fact that $\underline{x}_s^\top \mathbf{P}_2^\top \underline{x}_t = \underline{x}_t^\top \mathbf{P}_2 \underline{x}_s$ and obtain

$$\begin{aligned} f(\underline{x}_s, \underline{x}_t) &= \frac{1}{K(\mathbf{P})} \exp\left((\underline{x}_s^\top, \underline{x}_t^\top)\mathbf{P}(\underline{x}_s^\top, \underline{x}_t^\top)^\top\right) \\ &= \frac{1}{K(\mathbf{P})} \exp\left(\underline{x}_s \mathbf{P}_1 \underline{x}_s + 2\,\underline{x}_s^\top \mathbf{P}_2^\top \underline{x}_t + \underline{x}_t \mathbf{P}_3 \underline{x}_t\right) \end{aligned}$$

Completing the square using $\underline{x}_s \mathbf{P}_2^\top \mathbf{P}_3^{-1} \mathbf{P}_2 \underline{x}_s$ yields the desired result. □

With this representation, we obtain necessary and sufficient conditions on the matrix $\mathbf{P}$ in order to have a well-defined probability distribution. Thus, $\mathbf{P}_1$ needs to be symmetric, $\mathbf{P}_2$ may be arbitrary and $\mathbf{P}_3$ needs to be symmetric positive definite.

The formulation (6.8) immediately yields two further relationships with the Bingham distribution and the Gaussian distribution respectively. First,the Bingham distribution is the marginal distribution of $\underline{x}_s$, because marginalization yields

$$\begin{aligned} f(\underline{x}_s) &= \int_{\mathbb{R}^2} f(\underline{x}_s, \underline{x}_t)\ \mathrm{d}\underline{x}_t \\ &\propto \exp(\underline{x}_s^\top \mathbf{S}\, \underline{x}_s)\ . \end{aligned} \tag{6.9}$$

The second relationship is the fact that the density of $\underline{x}_t$ conditioned on a fixed $\underline{x}_s$ is Gaussian. This is easily seen from

$$\begin{aligned} f(\underline{x}_t \,|\, \underline{x}_s) &= \frac{f(\underline{x}_t, \underline{x}_s)}{f(\underline{x}_s)} \\ &\propto \exp\left((\underline{x}_t - \mathbf{A}\underline{x}_s)^\top \mathbf{P}_3 (\underline{x}_t - \mathbf{A}\underline{x}_s)\right)\ . \end{aligned}$$

Here $\propto$ refers to proportionality with respect to $\underline{x}_t$. From $\mathbf{P}_3^{-1} = -\frac{1}{2}(-\frac{1}{2}\mathbf{P}_3)^{-1}$, we have $\underline{x}_t|\underline{x}_s \sim \mathcal{N}(\mathbf{A}\underline{x}_s, -\frac{1}{2}\mathbf{P}_3^{-1})$.

These two relationships can be used to reduce the computation of a normalization constant of the new distribution to the computation of the normalization constant of the Bingham distribution.

Proposition 6.3. *The normalization constant $K(\mathbf{P})$ is given by*

$$K(\mathbf{P}) = 2\pi\sqrt{\det\left(-\frac{1}{2}\mathbf{P}_3^{-1}\right)} \cdot N(\mathbf{P}_1 - \mathbf{P}_2^\top \mathbf{P}_3^{-1} \mathbf{P}_2)^{-1}\ ,$$

where $N(\cdot)$ is the normalization constant of the Bingham distribution.

Proof. First, consider the integral over the unnormalized Gaussian

$$\int_{\mathbb{R}^2} \exp\left((\underline{x}_t - \mathbf{A}\underline{x}_s)^\top \mathbf{P}_3 (\underline{x}_t - \mathbf{A}\underline{x}_s)\right) \, \mathrm{d}\underline{x}_s = 2\pi \sqrt{\det\left(-\frac{1}{2}\mathbf{P}_3^{-1}\right)} \, . \quad (6.10)$$

From the fact, that $\underline{x}_s$ follows a Bingham distribution, we know its normalization constant to be given by $N(\mathbf{P}_1 - \mathbf{P}_2^\top \mathbf{P}_3^{-1} \mathbf{P}_2)$. Furthermore, from (6.9), we know that this Bingham normalization constant is a product of the Gaussian normalization constant given in (6.10) and $K(\mathbf{P})^{-1}$. Solving for $K(\mathbf{P})$ yields the desired result. □

Furthermore, a random sampling procedure can be derived from these two relationships. It is based on a two step approach. First, a random sample $\underline{x}_s$ is generated from a two dimensional Bingham distribution. Second, a Gaussian sample $\underline{x}_t$ is generated from $\mathcal{N}(\mathbf{A}\underline{x}_s, -\frac{1}{2}\mathbf{P}_3^{-1})$. Consequently, the entire resulting sample is then obtained as $(\underline{x}_s^\top, \underline{x}_t^\top)^\top$.

The representation in Proposition 6.2 is also useful for obtaining the modes of the proposed distribution. For the circular part $\underline{x}_s$, the modes are given by $\pm\underline{m}$, where $\underline{m}$ is the eigenvector corresponding to the largest eigenvalue of $\mathbf{S}$. Using the fact, that the proposed density conditioned on the circular part is Gaussian with mean $\mathbf{A}\underline{m}$, we can obtain the modes of the entire density as $\pm(\underline{m}^\top, (\mathbf{A}\underline{m})^\top)^\top$.

6.4.1 Parameter Estimation

The relationships discussed so far are not only useful for the computation of the normalization constant. They also play a key role in parameter estimation, which is carried out similarly to the approach in [75]. The main idea behind the parameter estimation procedure is estimating the Bingham distribution parameters from the first two entries of all provided samples in order to obtain $\mathbf{S}$ and then using multivariate linear regression [3] to obtain $\mathbf{A}$ (as defined in Proposition 6.2) and $\mathbf{P}_3$. This is sufficient to finally obtain an estimate of the entire parameter matrix $\mathbf{P}$.

Algorithm 6.1 $SE(2)$ Distribution Parameter Estimation

procedure ESTIMATESE2($\underline{s}_1, \ldots, \underline{s}_m$)
    $(\hat{\mathbf{M}}, \hat{\mathbf{Z}}) \leftarrow \text{EstimateBingham}((\underline{s}_1^{(1)}, \underline{s}_1^{(2)})^\top, \ldots, (\underline{s}_m^{(1)}, \underline{s}_m^{(2)})^\top)$;
    $\hat{\mathbf{S}} \leftarrow \hat{\mathbf{M}}\,\hat{\mathbf{Z}}\,\hat{\mathbf{M}}^\top$;
    $\mathbf{C}_1 \leftarrow \sum_{i=1}^m (\underline{s}_i^{(3)}, \underline{s}_i^{(4)})^\top \cdot (\underline{s}_i^{(1)}, \underline{s}_i^{(2)})$;
    $\mathbf{C}_2 \leftarrow \sum_{i=1}^m (\underline{s}_i^{(1)}, \underline{s}_i^{(2)})^\top \cdot (\underline{s}_i^{(1)}, \underline{s}_i^{(2)})$;
    $\hat{\mathbf{B}} \leftarrow \mathbf{C}_1 \mathbf{C}_2^{-1}$;
    $\hat{\mathbf{P}}_3 \leftarrow \left(\frac{2}{m} \sum_{i=1}^m \left(\begin{pmatrix} \underline{s}_i^{(3)} \\ \underline{s}_i^{(4)} \end{pmatrix} - \hat{\mathbf{B}} \begin{pmatrix} \underline{s}_i^{(1)} \\ \underline{s}_i^{(2)} \end{pmatrix} \right) \cdot \left(\begin{pmatrix} \underline{s}_i^{(3)} \\ \underline{s}_i^{(4)} \end{pmatrix} - \hat{\mathbf{B}} \begin{pmatrix} \underline{s}_i^{(1)} \\ \underline{s}_i^{(2)} \end{pmatrix} \right)^\top \right)^{-1}$
    $\hat{\mathbf{P}}_2 \leftarrow \hat{\mathbf{P}}_3\,\hat{\mathbf{B}}$;
    $\hat{\mathbf{P}}_1 \leftarrow \hat{\mathbf{S}} - \hat{\mathbf{P}}_2^\top \hat{\mathbf{P}}_3^{-1} \hat{\mathbf{P}}_2$;
    $\hat{\mathbf{P}} \leftarrow \begin{pmatrix} \hat{\mathbf{P}}_1 & \hat{\mathbf{P}}_2^\top \\ \hat{\mathbf{P}}_2 & -\hat{\mathbf{P}}_3 \end{pmatrix}$;
    return $\hat{\mathbf{P}}$;
end procedure

The procedure is shown in Algorithm 6.1. It produces an estimate of the distribution parameters from samples $\underline{s}_1, \ldots, \underline{s}_m \in \Omega$ which are assumed to be given in a form in which the first two entries $\underline{s}_i^{(1)}$, $\underline{s}_i^{(2)}$ denote the circular part and the second two entries $\underline{s}_i^{(3)}$, $\underline{s}_i^{(4)}$ denote the unconditioned part. Estimation of the Bingham part can be performed as proposed in Algorithm 5.1. Multivariate linear regression is carried out as described in [3, Theorem 8.2.1], that is

$$\hat{\mathbf{B}} = \mathbf{C}_1 \mathbf{C}_2^{-1} ,$$

where

$$\mathbf{C}_1 := \sum_{i=1}^m (\underline{s}_i^{(3)}, \underline{s}_i^{(4)})^\top \cdot (\underline{s}_i^{(1)}, \underline{s}_i^{(2)}) , \quad \mathbf{C}_2 := \sum_{i=1}^m (\underline{s}_i^{(1)}, \underline{s}_i^{(2)})^\top \cdot (\underline{s}_i^{(1)}, \underline{s}_i^{(2)}) .$$

And for the covariance

$$-\frac{1}{2}\hat{\mathbf{P}}_3^{-1} = \frac{1}{m} \sum_{i=1}^m \left(\begin{pmatrix} \underline{s}_i^{(3)} \\ \underline{s}_i^{(4)} \end{pmatrix} - \hat{\mathbf{B}} \begin{pmatrix} \underline{s}_i^{(1)} \\ \underline{s}_i^{(2)} \end{pmatrix} \right) \cdot \left(\begin{pmatrix} \underline{s}_i^{(3)} \\ \underline{s}_i^{(4)} \end{pmatrix} - \hat{\mathbf{B}} \begin{pmatrix} \underline{s}_i^{(1)} \\ \underline{s}_i^{(2)} \end{pmatrix} \right)^\top .$$

The entire estimation procedure can also be formulated in a more general way EstimateBingham$((p_1, \underline{s}_1), \ldots, (p_m, \underline{s}_m))$ where p_i denote probability weights in order to make consideration of weighted samples possible. This requires including the weights in the computation of $\mathbf{C}_1$, $\mathbf{C}_2$ and $\hat{\mathbf{P}}_3$. Besides that, the estimation procedure remains unchanged.

6.4.2 Representation of Uncertain Rigid-Body Motions

The newly derived distribution can be used to represent planar rigid-body motions. Its mode can be interpreted as a manifold equivalent of a mean rigid-body motion (a similar interpretation is also made in [75]). Thus, a suitable matrix needs to be chosen for obtaining a density with a certain predefined mean. This is done in the following way. Assume $\mathbf{P}$ is the parameter matrix of the newly proposed distribution with corresponding mode $(1, 0, 0, 0)$. Then, a distribution with desired mode $\underline{a}$ can be obtained as follows. Let $\mathbf{Q}_a$ denote the matrix representation of $\underline{a}^{-1}$. Using (6.4), it can be seen that

$$
\begin{aligned}
f(\underline{x}) &\propto \exp((\mathbf{D}\,\mathbf{Q}_a\,\mathbf{D}\,\underline{x})^\top\,\mathbf{P}\,(\mathbf{D}\,\mathbf{Q}_a\,\mathbf{D}\,\underline{x})) \\
&= \exp(\underline{x}^\top\,\mathbf{D}\,\mathbf{Q}_a^\top\,\mathbf{D}\,\mathbf{P}\,\mathbf{D}\,\mathbf{Q}_a\,\mathbf{D}\,\underline{x}) \ ,
\end{aligned}
$$

with $\mathbf{D} = \operatorname{diag}(1, -1, -1, -1)$ and thus $\mathbf{D}^\top = \mathbf{D}$. An analogous argument using (6.5) establishes that if $\underline{x}$ follows the proposed distribution with parameter matrix $\mathbf{P}$, then $\underline{x}^{-1}$ is also distributed according to this distribution with parameter matrix $\mathbf{D}\,\mathbf{P}\,\mathbf{D}$.

6.5 Deterministic Sampling

In this section, a deterministic sampling method is proposed, which serves as a discrete approximation of the distribution presented above. The following approach differs on a principal level from the sampling approaches discussed in both preceding chapters. First, it differs from the sampling of the Bingham distribution, which is a purely moment-based approach and thus could in theory be reasonably applied to every antipodally symmetric distribution on the hypersphere. Second, the proposed sampling is not a shape approximation as discussed in chapter 4.

Algorithm 6.2 $SE(2)$ Distribution Deterministic Sampling

procedure DETERMINISTICSAMPLING($\mathbf{P}_1$, $\mathbf{P}_2$, $\mathbf{P}_3$)
  $\mathbf{S} \leftarrow \mathbf{P}_1 - \mathbf{P}_2^\top \mathbf{P}_3^{-1} \mathbf{P}_2$;
  $\mathbf{A} \leftarrow -\mathbf{P}_3^{-1} \mathbf{P}_2$;
  $(\underline{v}_1, \underline{v}_2, \lambda_1, \lambda_2) \leftarrow$ Eigendecomposition($\mathbf{S}$) ▷ Assuming $\lambda_i \leq \lambda_{i+1}$
  $\mathbf{M} \leftarrow (\underline{v}_1, \underline{v}_2)$;
  $\mathbf{Z} \leftarrow \mathrm{diag}(\lambda_1 - \lambda_2, 0)$;
  $(\underline{a}_1, \ldots, \underline{a}_N, w_{s,1}, \ldots, w_{s,N}) \leftarrow$ SampleBingham($\mathbf{M}$,$\mathbf{Z}$)
  $(\underline{b}_1, \ldots, \underline{b}_M, w_{t,1}, \ldots, w_{t,M}) \leftarrow$ SampleGaussian($\underline{0}$, $-\frac{1}{2}\mathbf{P}_3$)
  $k \leftarrow 1$;
  for $i \in \{1, ..., N\}$ **do**
    for $j \in \{1, ..., M\}$ **do**
      $\underline{x}_k \leftarrow \begin{pmatrix} \underline{a}_i \\ \underline{b}_j + \mathbf{A} \cdot \underline{a}_i \end{pmatrix}$;
      $p_k \leftarrow w_{s,i} \cdot w_{t,j}$;
      $k \leftarrow k + 1$;
    end for
  end for
  return $(p_1, \underline{x}_1), \ldots, (p_{N \cdot M}, \underline{x}_{N \cdot M})$;
end procedure

The deterministic sampling scheme proposed here can be seen as a hybrid approach, which can be more generally seen as an entire class of sampling approaches. The key idea is once again to make use of the fact that the marginal distribution of the circular part is Bingham and conditioning the non-restricted part on a given circular part yields a Gaussian. First, we use a deterministic sampling scheme to obtain the approximation of the Bingham distribution. This yields m_s samples $\underline{m}_1, \ldots, \underline{m}_{m_s}$. Second, for each of these samples $\underline{m}_i$ we obtain a deterministic sample set approximating a $\mathcal{N}(\mathbf{A}\underline{m}_i, -\frac{1}{2}\mathbf{P}_3)$ distribution using m_t samples. This is brought together in a resulting sample set consisting of $m = m_s \cdot m_t$ samples.

Thus, the deterministic sampling procedure is similar to random sampling. It is hybrid in the sense that, both involved sampling procedures can be purely moment based. However, the entire procedure still makes use of the particular structure of the distribution and is thus not easily generalized to other probability distributions on $SE(2)$.

It does not matter which deterministic sampling procedure is used for approximating the Gaussian density and the Bingham density respectively. Furthermore, it suffices to sample only one Gaussian with covariance $-\frac{1}{2}\mathbf{P}_3$. A suitable repositioning operation is then used to obtain the desired mean.

The entire sampling scheme is shown in Algorithm 6.2. The methods SampleBingham, and SampleGaussian refer to a deterministic sampling of the Bingham and the Gaussian distributions respectively.

6.6 A Filter for Rigid-Body Motions in the Plane

We consider a general system model, where an uncertain system state is represented by a dual-quaternion. Moreover, all arising noise is also modeled using the newly proposed distribution. That is, our system model is given by

$$\underline{x}_{t+1} = a(\underline{x}_t, \underline{w}_t) ,$$

where $\underline{x}_t$ is characterized using the proposed distribution with parameter matrix $\mathbf{P}^e_{t+1}$ and $\underline{w}_t$ is assumed to be distributed according to the proposed distribution with parameter matrix $\mathbf{P}_w$. Furthermore, we require the system function $a(\cdot,\cdot)$ to respect antipodal symmetry in both parameters. That is

$$a(-\underline{x}, \underline{w}) = -a(\underline{x}, \underline{w}) , \qquad a(\underline{x}, -\underline{w}) = -a(\underline{x}, \underline{w}) .$$

In our measurement model, we consider noisy direct measurements, i.e., the model is given by

$$\underline{z}_t = \underline{x}_t \boxplus \underline{v}_t , \tag{6.11}$$

where $\underline{v}_t$ is following the proposed distribution with parameter matrix $\mathbf{P}_v$. This model not only accounts for a manifold equivalent of zero-mean noise, but also makes considerations of biased measurement models possible as long as the bias is independent from the current system state, and thus, can be modeled as a suitably chosen mode of the noise distribution, i.e., a suitably chosen $\mathbf{P}_v$.

6.6.1 Prediction

The prediction is entirely based on a twofold deterministic sampling procedure and requires three approximations. First, $g(\underline{x}_t, \underline{w}_t)$ is in general not distributed according to the proposed distribution. This is also true for the special case $\underline{x}_t \boxplus \underline{w}_t$. Thus, we will approximate the distribution of $g(\underline{x}_t, \underline{w}_t)$ by the proposed distribution. A first approach would be using the method of moments. From the fact that g respects antipodal symmetry, it follows that $E(g(\underline{x}, \underline{w})) = 0$. Thus, we could match the second moment, as in the Bingham case. This would require the computation of $\mathbb{E}(g(\underline{x}, \underline{w}) \cdot g(\underline{x}, \underline{w})^\top)$, which is computationally burdensome. The reason for this is the fact, that currently the only way to carry out this computation would be direct numerical integration involving a triple integral. Two further approximations are introduced to avoid this computational burden. These are the deterministically obtained sample based approximations (according to the method proposed in the previous section) of the densities of the current estimate $\underline{x}_t^e$ (denoted as $\underline{s}_{x,i}$ with weights $p_{x,i}$ and $i \in \{1, \dots m\}$) and the system noise $\underline{w}_t$ (denoted as $\underline{s}_{w,i}$ with weights $p_{w,i}$ and $i \in \{1, \dots m\}$). These approximations are used to compute

$$\underline{s}_{m\cdot(i-1)+j} = a(\underline{s}_i, \underline{w}_j) \ ,$$
$$p_{m\cdot(i-1)+j} = p_{x,i} \cdot p_{w,j} \ .$$

Finally, the parameter estimation procedure proposed above is used to obtain the parameters $\hat{\mathbf{P}}_{t+1}^p$ for the predicted density. The entire prediction step is shown in Algorithm 6.3 and visualized in Figure 6.1. The point estimate $\hat{\underline{x}}_{t+1}^p$ is obtained as the mode of the distribution computed in this prediction.

6.6.2 Measurement Update

The proposed filtering procedure makes a closed-form measurement update possible. It starts off with the predicted estimate given as the parameter matrix $\mathbf{P}_t^p$. First, we reformulate the measurement model (6.11) as

$$\underline{x}_t^{-1} \boxplus \underline{z}_t = \underline{v}_t \ .$$

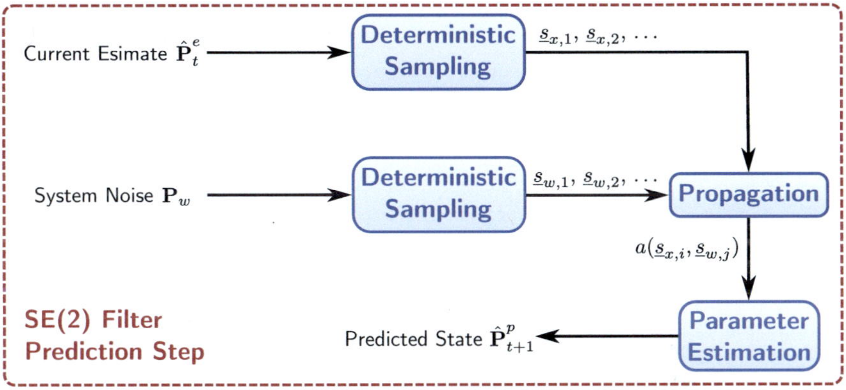

Figure 6.1.: Prediction step of SE(2) filter using deterministic sampling.

Using Bayes theorem, we obtain

$$f(\underline{x}_t|\underline{z}_t) \propto f_{\underline{v}}(\underline{x}_t^{-1} \boxplus \underline{z}_t) \cdot f(\underline{x}_t) \ .$$

The prior $f(\underline{x}_t)$ is known, and thus, the main challenge is expressing the likelihood in terms of $\underline{x}_t$. This is carried out by making use of (6.5) and (6.4), and thus obtaining

$$f_{\underline{v}}(\underline{x}_t^{-1} \boxplus \underline{z}_t) \overset{(6.5)}{=} f_{\underline{v}}(\mathbf{D}(\underline{z}_t^{-1} \boxplus \underline{x}_t)) \overset{(6.4)}{=} f_{\underline{v}}(\mathbf{D}\,\mathbf{D}\,\mathbf{Z}\,\mathbf{D}\,\underline{x}) \ ,$$

where $\mathbf{D} = \operatorname{diag}(1, -1, -1, -1)$ and $\mathbf{Z}$ is the matrix representation of the dual quaternion $\underline{z}^{-1}$. As a consequence of $\mathbf{DD} = \mathbf{I}$, the above identity simplifies to

$$f_{\underline{v}}(\underline{x}_t^{-1} \boxplus \underline{z}_t) = f(\mathbf{Z}\,\mathbf{D}\,\underline{x}) \ .$$

Finally, we obtain the entire resulting posterior density as

$$f(\underline{x}_t|\underline{z}_t) \propto \exp\left(\underline{x}^\top \left(\mathbf{D}\,\mathbf{Z}^\top\,\mathbf{P}_v\,\mathbf{Z}\,\mathbf{D} + \mathbf{P}_t^p\right) \underline{x}\right) \ .$$

The entire resulting algorithm is shown in Algorithm 6.4. There, the function MatrixRepresentation($\underline{z}$) denotes the matrix representation of the dual quaternion $\underline{z}$.

Algorithm 6.3 SE(2) Filter Prediction Step

procedure SE2PREDICT($\mathbf{P}_t^e$, $\mathbf{P}_w$)
  $(p_{x,1}, \underline{s}_{x,1}), \dots, (p_{x,m}, \underline{s}_{x,m}) \leftarrow$ SE2DeterministicSampling($\mathbf{P}_t^e$)
  $(p_{w,1}, \underline{s}_{w,1}), \dots, (p_{w,m}, \underline{s}_{w,m}) \leftarrow$ SE2DeterministicSampling($\mathbf{P}_w$)
  $k \leftarrow 1$;
  for $i \in \{1, \dots, m\}$ **do**
    for $j \in \{1, \dots, m\}$ **do**
      $p_k \leftarrow p_{x,i} \cdot p_{w,j}$;
      $\underline{s}_k \leftarrow a(\underline{s}_{x,i}, \underline{s}_{w,j})$;
      $k \leftarrow k + 1$;
    end for
  end for
  $\mathbf{P}_{t+1}^p \leftarrow$ EstimateSE2($(p_1, \underline{s}_1), \dots, (p_m, \underline{s}_m)$)
  return $\mathbf{P}_{t+1}^p$;
end procedure

Algorithm 6.4 SE(2) Filter Update

procedure MEASUREMENTUPDATE($\mathbf{P}_t^p$, $\mathbf{P}_v$, $\underline{z}$)
  $\mathbf{Z} \leftarrow$ MatrixRepresentation($\underline{z}^{-1}$);
  $\mathbf{D} \leftarrow \mathrm{diag}(1, -1, -1, -1)$;
  $\mathbf{P}_t^e \leftarrow \mathbf{D}\,\mathbf{Z}^\top\,\mathbf{P}_v\,\mathbf{Z}\,\mathbf{D} + \mathbf{P}_t^p$;
  return $\mathbf{P}_t^e$;
end procedure

6.7 Evaluations

All evaluations were carried out using Matlab 2014b. The deterministic sampling was implemented using the algorithm from the previous chapter for sampling the circular part (with $\lambda = 0.5$) and a naïve implementation of the UKF sampling with equal weights for the Gaussian part. The correct covariance of the Gaussian samples was obtained by multiplying the samples of a standard Gaussian with the Cholesky decomposition of the desired covariance matrix.

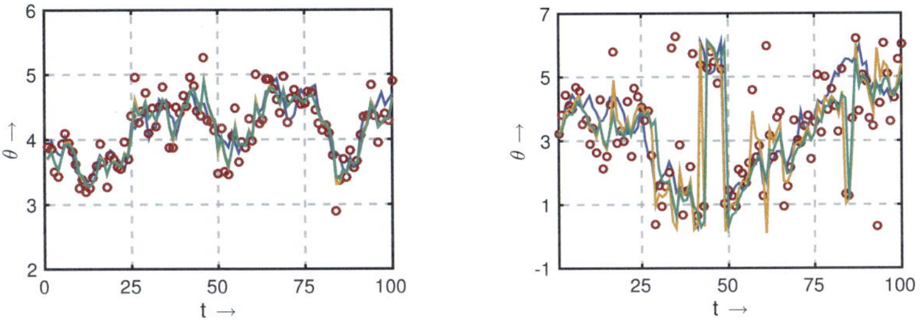

(a) Orientation.

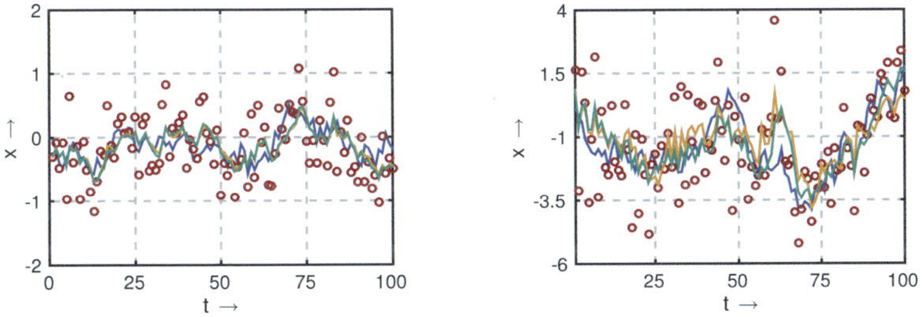

(b) x-Position.

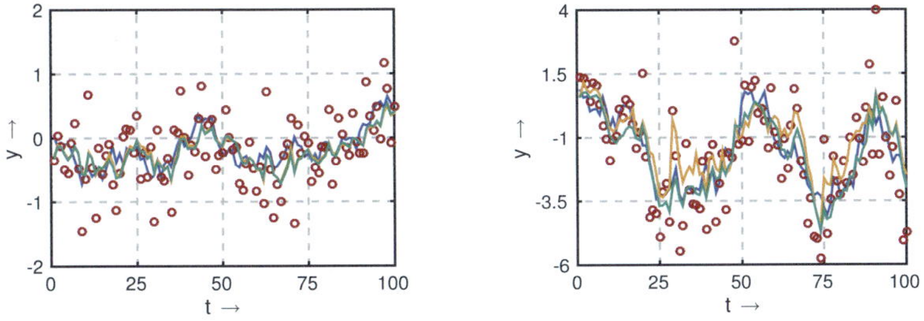

(c) y-Position.

Figure 6.2.: A low noise (let column) and a high noise (right column) simulation run showing the true system evolution (blue lines), its estimate (green lines), the UKF estimate (orange), and the measurements (red circles).

For our simulation run, we assumed a noisy identity model

$$\underline{x}_{t+1} = \underline{x}_t \boxplus \underline{w}_t$$

and a direct measurement model

$$\underline{y}_t = \underline{x}_t \boxplus \underline{v}_t$$

This model can also account for control inputs by choosing a suitable parameter matrix for the distribution of $\underline{w}_t$. Furthermore, it is also possible to compute the second moment of $\underline{x}_{t+1}$ from the second moments of $\underline{x}_t$ and $\underline{w}_t$ using moment matching. However, for our simulation we are interested in the approach based on the proposed deterministic sampling scheme.

A manifold variant of the UKF was used as comparison, which was also based on dual quaternions for representing quantities on the $SE(2)$. Therefore, its implementation required some adoptions in order to make it work for the considered estimation problem. First, we project the predicted state and the updated estimate back to the manifold in order to obtain a feasible result. Second, an intelligent repositioning of measurements is introduced in order to account for antipodal symmetry. This is done by comparing the obtained measurement $\underline{z}_t$ with the expected measurement $\hat{\underline{z}}_t$ (in case of the manifold equivalent of zero noise, this corresponds to the current estimate) and multiplying the current estimate with -1 if the resulting value is closer to the expected measurement.

The proposed distribution was used in order to generate the ground truth. In order to adapt the UKF to this uncertainty, a manifold aware moment matching procedure was performed. That is, random sampling was used to compute the covariance of all noise terms involved. For this procedure, it is infeasible to simply compute $\mathbf{Cov}(\underline{w}_t)$, because this would not consider the intelligent repositioning of the measurements. Thus, it is necessary to compute the covariance matrix of $r(\underline{w}_t)$, where $r(\cdot)$ is a function repositioning w on the same side as one of its modes.

Two typical filter runs with different noise levels were carried out. For the first run, the initial value $\underline{x}_0$ was generated using the proposed distribution with parameter matrix $\mathbf{P}_0 = \text{diag}(-10, 0, 1, 1)$. The system noise was modeled as $\mathbf{P}_w = \text{diag}(0 - 55 - 100 - 100)$ and the measurement noise as $\mathbf{P}_v = \mathrm{diag}(0, -30, -10, -10)$. Higher noise levels and dependency between the angular and the linear quantities were introduced in the second run. This was achieved by using the following parameter matrices $\mathbf{P}_0 = \mathrm{diag}(-10, 0, -1, -1)$, $\mathbf{P}_w = \mathrm{diag}(0, -3, -1, -1)$, and

$$\mathbf{P}_v = \begin{pmatrix} 0 & 0 & 0 & 1 \\ 0 & -15 & 0 & 0 \\ 0 & 0 & -10 & 0 \\ 1 & 0 & 0 & -10 \end{pmatrix} .$$

The results of both simulation runs are visualized in Figure 6.2. In order to provide some intuition, this visualization is based on reconstructing the original rigid-body motions from the dual quaternion representation. That is, it shows the rotation and subsequent translations as represented by the underlying dual quaternions. The results look similar to other directional estimation problems and are in line with our expectation, that small uncertainties can be handled by filters assuming linear underlying state spaces. As the uncertainty grows, these filters (in the particular example here, the UKF) fail to capture the geometric structure of the underlying manifold.

The evaluation was performed using 100 runs in the low noise scenario. Its results are visualized in Figure 6.3. The proposed filter is slightly better in terms of angular error in this scenario. This part is similar to the Bingham filter. However, the angular error has a considerable impact on the error of the accuracy of the estimated position. This is due to the fact that the dual-quaternion representation of a rigid-body motion links the angular part with the translational part. That is, an error in the angular part has also some effect in the estimated position. Due to a correct consideration of the manifold structure within our distribution, this relationship is taken into account. Thus, the newly proposed method outperforms filters based on assuming a Gaussian distribution.

6.8 Summary and Discussion

This chapter proposed a novel filter for the representation of uncertain rigid-body motions in the plane. It is based on two contributions. First, a novel distribution from the exponential family of probability distributions

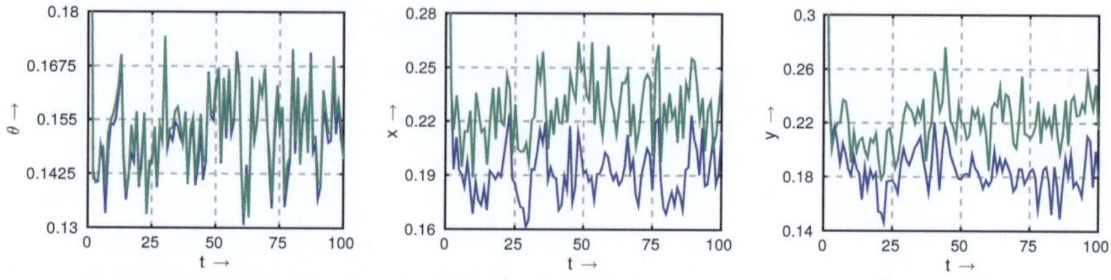

(a) Orientation **(b)** x-Position **(c)** y-Position

Figure 6.3.: Mean error after 100 runs in the low noise scenario showing the proposed approach (blue) and the adapted UKF (green).

was presented, making the representation of planar uncertain rigid-body motions easily possible by using a multiplicative subgroup of unit dual quaternions. This distribution makes a closed-form Bayes update possible and is directly related to the Bingham distribution.

Consequently, a number of problems of typical approaches in dynamic state estimation on nonlinear domains is avoided. These problems are typically a result of wrongly assuming linearity of the underlying state-space (as done when using some variant of nonlinear Kalman filters) or by performing approximations in order to ensure that prior and posterior distribution are of the same type (as done in approaches based on densities defined on the tangent space, e.g., [27]).

The second contribution is a novel deterministic sampling scheme, which is based on a hybrid approach that decomposes the problem of deterministically sampling the newly proposed distribution into the problems of deterministically sampling Bingham and Gaussian distributions. This approach is highly adaptable, because arbitrary underlying sampling schemes for the Bingham and the Gaussian may be used. However, its drawback is that it requires a fairly high number of samples. On the other hand, this property might be desirable when large nonlinearities are involved. Both contributions give rise to a novel filtering scheme truly respecting the structure of the underlying manifold and thus outperforming approaches based on assuming linearity.

CHAPTER 7

Conclusions and Outlook

Problems in dynamic state estimation involve describing system and measurement models, quantifying and understanding the nature of the uncertainties, deriving stochastic estimators, and implementing these estimators in a computationally tractable way. Typically, some sort of approximation has to be made in order to overcome each of these challenges. These approximations might involve time-discretization, use of an approximate model, or simplifying assumptions about uncertainty of the system. The goal of these approximations is to provide computationally feasible algorithms while maintaining sufficiently robust and accurate results. This thesis focused on deterministic approximation of uncertainties and consideration of two important nonlinear domains, namely the domain of orientations and the domain of planar rigid-body motions.

7.1 Key Contributions

A Metric for Probability Distributions

In this thesis, a L_2 distance was used for comparing probability distributions. It is capable of simultaneously considering continuous and discrete probability distributions. This is made possible by using localized cumulative distributions (LCDs), a representation of local probability mass around a given point. The contribution to the further development of this measure was twofold. First, we have discussed the setup in which this measure is an actual metric and considered the special case based on Gaussian kernels. Second, we have shown that the transform used to obtain an LCD representation (when using Gaussian-RBF kernels) is directly related to the Weierstrass transform. Thus, results from the theory on Weierstrass transforms are applicable to the LCD representation.

Deterministic Sampling of the Gaussian Distribution

The LCD-based metric was used to derive a shape approximation of Gaussian densities (and more generally of Gaussian mixture densities). This approximation requires a numerical procedure that places the samples in a way minimizing the distance measure. The burden of both, numerically evaluating the distance measure and performing numerical opimization based on this measure, was addressed by deriving a computationally more efficient representation for the case of approximating even-dimensional Gaussians. Furthermore, for use in real-time filtering applications, we discussed the application of a scheme based on using a precomputed approximation of the standard normal distribution. We have also shown that, when applying the distance measure in a scenario with equal means, it is not necessary to choose a maximum kernel width. That is, in this case, the measure can be evaluated in a way that considers every possible kernel size. Finally, our evaluations have shown that the use of proposed approximation for propagation of uncertainties outperforms state-of-the-art approximate integration approaches, either in terms of computational complexity or in terms of accuracy.

Sample-based Filters on Nonlinear Domains

This work considers filtering on nonlinear domains by providing filtering techniques based on probability distributions which are truly defined on these domains and thus reflect their geometric structure. The domains particularly considered are the hypersphere and the cartesian product of a circle and the Euclidean plane. The hypersphere is of our interest, because the four dimensional special case can be used to represent uncertain orientation. Whereas, the cartesian product of the circle and the Euclidean space is of particular interest for representing rigid-body motions on the plane. On the hypersphere, we make use of the Bingham distribution. Its antipodal symmetry comes as a useful property that accounts for the fact that unit quaternions are a double cover of the group of orientations. As there was no counterpart for planar rigid-body motions, we proposed a new Bingham-like distribution for representing uncertainty on that group. In the proposed setting, unit dual quaternions were used as representation of elements of $SE(2)$. Similarly, to the case of uncertain orientations, unit quaternions are a double cover of the $SE(2)$. Both distributions were used to derive sample-based stochastic filters. These filters can be thought of as the manifold-equivalent to the linear regression Kalman filter.

Deterministic Sampling on Nonlinear Domains

In order to avoid the use of possibly costly random sampling for our filtering techniques on nonlinear domains, we propose deterministic sampling techniques on these domains. In the hyperspherical case, our sampling technique is based on moment matching. Thus, it has a very general application, because it is not limited to approximating the Bingham distribution, but can be used to approximate arbitrary antipodally symmetric distributions on arbitrary-dimensional hyperspheres. In that sense, the proposed sampling technique is an antipodally symmetric hyperspherical analogue of the unscented transform. A distribution-dependent approach was taken for a sample-based approximation of the newly proposed distribution for the representation of planar rigid-body motions. The proposed approach considers the probabilistic structure within the distribution. In particular, we make use of the fact that the marginal distribution of the circular part is a (circular) Bingham distribution whereas the distribution of the non-circular part conditioned on a given value of the circular part is Gaussian. This gives rise to a hybrid sampling approach, which uses deterministic sampling of both, sampling a Bingham and a Gaussian distribution. In a practical implementation, arbitrary sampling procedures may be used. It is even possible to replace one of the procedures by a random sampling technique while keeping the other deterministic.

7.2 Relevance

State-of-the-art sample-based filters used either a fixed number of samples, a very fast (usually at least quadratically in the dimension) growing number of samples, or involved random sampling, and thus, potentially suffered from low flexibility, high computational burden, or quality loss. Furthermore, the alternative of using adaptive numerical integration techniques also suffers from a high computational demand that makes it unsuitable for many real-time dynamic state estimation scenarios. This thesis overcomes these limitations by proposing a highly adaptable sampling scheme of the Gaussian distribution. Even though offlien precomputation imposes some suboptimality, the resulting density approximation can be chosen in a way ensuring a desired quality level for the propagation task. Thus, this methodology combines the advantages of deterministic computations with the flexibility of choosing an arbitrary number of samples without the use of randomized methods. Furthermore, the systematically correct consideration of nonlinear domains in nonlinear filtering facilitates simultaneous use of good and poor sensor data within the same application. That is, the proposed algorithms avoid throwing away measurements on nonlinear domains, and thus, improve the efficiency of technical sensor systems.

7.3 Future Research

Dependencies Between Quantities on Nonlinear Domains
The consideration of dependencies is already a challenging task on linear state spaces. Due to the nature of filters that make a Gaussian assumption, it is usually assumed that dependencies arise as a linear relationship that can be adequately described by the covariance. However, this is not necessarily the case. E.g., when a nonlinear measurement model is employed, the dependency structure between the uncertain state and its measurement might be highly nonlinear. Assuming a linear relationship might be still a sufficient approximation in many cases. In adapted form, these considerations also hold for nonlinear domains. However, there is additionally another basic challenge for these domains. There is no canonical measure of dependency on nonlinear domains, and thus, even for the circular case there are several measures proposing some circular adaption of the covariance. A first approach for the circular case, considering the dependency between two circular quantities, was presented in [142]. A future challenge is the generalization of this approach to a higher number of dependent circular quantities, and dependencies on other nonlinear domains.

Distributed Filtering on Nonlinear Domains

So far, our filtering algorithms have assumed a very centralized setting. The presented algorithms were suitable for fusing estimates and measurements as long as no common process noise was involved. In the linear case, there has been a great body of research considering the problem of common process noise, e.g., [100], [99], [16], [84]. There has also been some research on this problem involving nonlinear system dynamics. However, to the best of the authors knowledge there is no systematic consideration of nonlinear domains for decentralized filtering.

Further Applications

This thesis mostly focused on deriving results contributing to the theory of stochastic filtering. As already mentioned, these contributions were motivated by real-world problems and the algorithms proposed within this work can be directly used as a replacement for existing nonlinear propagation algorithms or estimation methods. However, it is of considerable interest to investigate the applicability of the proposed results beyond the framework of dynamic state estimation, e.g., in the broader field of computational perception. This may involve object detection or stochastic control involving directional quantities.

APPENDIX A

Auxiliary Results

A.1 Exponential Integral

The exponential integral is defined as

$$\mathrm{Ei}(x) = \int_{-\infty}^{x} \frac{\mathrm{e}^y}{y} \,\mathrm{d}y \; . \tag{A.1}$$

The integral is understood in the sense of a Cauchy principal value for $x > 0$ due to divergence of the integrand at $y = 0$. A useful expansion of this integral is given by

$$\mathrm{Ei}(x) = \gamma + \ln|x| + \sum_{k=1}^{\infty} \frac{x^k}{k!\,k} \; , \tag{A.2}$$

where $\gamma \approx 0.5772$ is the Euler-Mascheroni constant. It is a simple consequence of equations 6.2.7 and 6.6.4 in [96].

A.2 Integration on S^n

The goal of this section is to explain how the notation we use in this thesis for integration on the hypersphere is understood.

For integration over the surface of unit hyperspheres, we use the notation

$$\int_{S^{n-1}} f(\underline{x}) \,\mathrm{d}\underline{x} \; . \tag{A.3}$$

In order to explain this notation, we introduce hyperspherical coordinates (with fixed radius 1). That is, we introduce variables $\varphi_1, \ldots, \varphi_{n-1}$ with $\varphi_1, \ldots, \varphi_{n-2} \in [0, \pi]$ and $\varphi_{n-1} \in [0, 2\pi)$. This is used in order to perform the following substitution

$$\begin{aligned} x^{(1)} &= \cos(\varphi_1) \ , \\ x^{(2)} &= \sin(\varphi_1) \cdot \cos(\varphi_2) \ , \\ &\vdots \\ x^{(n)} &= \sin(\varphi_1) \cdot \ldots \cdot \sin(\varphi_{n-2}) \cdot \cos(\varphi_{n-1}) \end{aligned}$$

This is used to define $\underline{x}(\varphi_1, \ldots, \varphi_{n-1})$. Now, the integral in (A.3) can be rewritten as follows

$$\begin{aligned} &\int_{S^{n-1}} f(\underline{x}) \, \mathrm{d}\underline{x} \\ &= \int_0^{2\pi} \underbrace{\int_0^{\pi} \cdots \int_0^{\pi}}_{n-2 \text{ times}} f(\underline{x}(\varphi_1, ..., \varphi_{n-1})) \\ &\qquad\qquad \times \left(\prod_{i=1}^{n-2} \sin(\varphi_i)^{n-1-i} \right) \mathrm{d}\varphi_1 \ldots \, \mathrm{d}\varphi_{n-1} \end{aligned}$$

The additional $\sin(\varphi_i)$ terms are a consequence of the transformation theorem. This choice ensures that our notation follows the intuition, that the integral of 1 over S^{n-1} yields the surface area of S^{n-1}.

List of Figures

List of Algorithms

Bibliography

[1] L. Aggoun and R. Elliot. *Measure Theory and Filtering.* Cambridge University Press, 2004.

[2] D. Alspach and H. Sorenson. Nonlinear Bayesian Estimation Using Gaussian Sum Approximations. *IEEE Transactions on Automatic Control*, 17(4):439–448, Aug. 1972.

[3] T. W. Anderson. *An Introduction to Multivariate Statistical Analysis.* Wiley-Interscience, 2003.

[4] M. Arulampalam, S. Maskell, N. Gordon, and T. Clapp. A Tutorial on Particle Filters for Online Nonlinear/Non-Gaussian Bayesian Tracking. *IEEE Transactions on Signal Processing*, 50(2):174 –188, 2002.

[5] M. Azmani, S. Reboul, J.-B. Choquel, and M. Benjelloun. A Recursive Fusion Filter for Angular Data. In *International Conference on Robotics and Biometrics (ROBIO)*, pages 882 –887, Dec. 2009.

[6] T. Bailey, J. Nieto, J. Guivant, M. Stevens, and E. Nebot. Consistency of the EKF-SLAM Algorithm. In *2006 IEEE/RSJ International Conference on Intelligent Robots and Systems*, pages 3562–3568, Beijing, China, 2006.

[7] A. Bain and D. Crisan. *Fundamentals of Stochastic Filtering.* Springer, 2009.

[8] E. Batschelet. *Circular Statistics in Biology.* Academic Press, 1981.

[9] C. Bingham. An Antipodally Symmetric Distribution on the Sphere. *The Annals of Statistics*, 2(6):1201–1225, Nov. 1974.

[10] A. Bishop. Information Fusion via the Wasserstein Barycenter in the Space of Probability Measures: Direct Fusion of Empirical Measures and Gaussian Fusion with Unknown Correlation, 2014.

[11] J. Bookshire and S. Teller. Extrinsic Calibration from Per-Sensor Egomotion. In *Robotics: Science and Systems*, Sydney, Australia, 2012.

[12] R. Booton. The Analysis of Nonlinear Control Systems with Random Inputs. In *Proceedings of the Symposium on Nonlinear Circuit Analysis*, New York, NY, USA, 1953.

[13] R. Booton. Nonlinear Control Systems with Random Inputs. *IRE Transactions on Circuit Theory*, CT-1(1), 1954.

[14] T. Brehard and J.-P. Le Cadre. Hierarchical particle filter for bearings-only tracking. *IEEE Transactions on Aerospace and Electronic Systems*, 43(4):1567–1585, Oct. 2007.

[15] C. Brunsdon and J. Corcoran. Using Circular Statistics to Analyse Time Patterns in Crime Incidence. *Computers, Environment and Urban Systems*, 30(3):300–319, May 2006.

[16] N. Carlson. Federated Square Root Filter for Decentralized Parallel Processors. *IEEE Transactions on Aerospace and Electronic Systems*, 26(3):517–525, May 1990.

[17] J. A. Carta, C. Bueno, and P. Ramírez. Statistical modelling of directional wind speeds using mixtures of von Mises distributions: Case study. *Energy Conversion and Management*, 49(5):897–907, May 2008.

[18] A. Chiuso and G. Picci. Visual Tracking of Points as Estimation on the Unit Sphere. In *The Confluence of Vision and Control*, pages 90–105. 1998.

[19] C. Chlebek, A. Hekler, and U. D. Hanebeck. Stochastic Nonlinear Model Predictive Control Based on Progressive Density Simplification. In *Proceedings of the 51st IEEE Conference on Decision and Control (CDC 2012)*, Maui, Hawaii, USA, Dec. 2012.

[20] W. K. Clifford. Preliminary Sketch of Biquaternions. *Proceedings of the London Mathematical Society*, s1-4(1):381–395, 1873.

[21] W. J. Cody and H. C. Thacher. Chebyshev Approximations for the Exponential Integral Ei(x). *Mathematics of Computation*, 23(106):289–289, May 1969.

[22] H. E. Daniels. Saddlepoint Approximations in Statistics. *The Annals of Mathematical Statistics*, 25(4):631–650, Dec. 1954.

[23] A. Doucet, S. Godsill, and C. Andrieu. On Sequential Monte Carlo Sampling Methods for Bayesian Filtering. *Statistics and Computing*, 10(3):197–208, July 2000.

[24] R. M. Dudley. *Real Analysis and Probability.* 2002.

[25] D. B. Duncan and S. D. Horn. Linear Dynamic Recursive Estimation from the Viewpoint of Regression Analysis. *Journal of the American Statistical Association*, 67(340):815–821, 1972.

[26] J. Dunik, O. Straka, and M. Simandl. Stochastic Integration Filter. *IEEE Transactions on Automatic Control*, 58(6):1561–1566, June 2013.

[27] W. Feiten, P. Atwal, R. Eidenberger, and T. Grundmann. 6D Pose Uncertainty in Robotic Perception. In T. Kröger and F. M. Wahl, editors, *Advances in Robotics Research*, pages 89–98. Springer Berlin Heidelberg, 2009.

[28] W. Feiten, M. Lang, and S. Hirche. Rigid Motion Estimation Using Mixtures of Projected Gaussians. In *Proceedings of the 16th International Conference on Information Fusion (Fusion 2013)*, pages 1465–1472, 2013.

[29] C. F. Gauß. *Theoria Combinationis Observationum Erroribus Minimis Obnoxiae.* 1823.

[30] C. F. Gauß. Supplementum Theoriae Combinationis Observationum Erroribus Minimis Obnoxiae. 1826.

[31] A. Gelb, J. F. Kasper, R. A. Nash, C. F. Price, and A. A. Sutherland. *Applied Optimal Estimation.* The MIT Press, 1974.

[32] A. L. Gibbs and F. E. Su. On Choosing and Bounding Probability Metrics. *International Statistical Review*, 70(3):419–435, Dec. 2002.

[33] J. Gill and D. Hangartner. Circular Data in Political Science and How to Handle It. *Political Analysis*, 18(3):316–336, June 2010.

[34] J. Glover. Bingham Statistics Library, 2012.

[35] J. Glover. *The Quaternion Bingham Distribution, 3D Object Detection, and Dynamic Manipulation.* Phd thesis, Massachusetts Institute of Technology, 2014.

[36] J. Glover and L. P. Kaelbling. Tracking 3-D Rotations with the Quaternion Bingham Filter. Technical report, MIT CSAIL, Mar. 2013.

[37] J. Glover, R. Rusu, and G. Bradski. Monte Carlo Pose Estimation with Quaternion Kernels and the Bingham Distribution. In *Proceedings of Robotics: Science and Systems*, Los Angeles, CA, USA, June 2011.

[38] J. S. J. Goddard. *Pose and Motion Estimation from Vision Using Dual Quaternion-based Extended Kalman Filtering.* PhD thesis, University of Tennessee, Jan. 1997.

[39] N. Gordon, D. Salmond, and A. Smith. Novel Approach to Nonlinear/Non-Gaussian Bayesian State Estimation. *IEE Proceedings F Radar and Signal Processing*, 140(2):107, Apr. 1993.

[40] C. Goutis and G. Casella. Explaining the Saddlepoint Approximation. *The American Statistician*, 53(3):216–224, Feb. 1999.

[41] U. D. Hanebeck. PGF 42: Progressive Gaussian Filtering with a Twist. In *Proceedings of the 16th International Conference on Information Fusion (Fusion 2013)*, Istanbul, Turkey, July 2013.

[42] U. D. Hanebeck. Optimal Reduction of Multivariate Dirac Mixture Densities. *arXiv preprint: Systems and Control (cs.SY)*, Nov. 2014.

[43] U. D. Hanebeck, M. F. Huber, and V. Klumpp. Dirac Mixture Approximation of Multivariate Gaussian Densities. In *Proceedings of the 2009 IEEE Conference on Decision and Control (CDC 2009)*, Shanghai, China, Dec. 2009.

[44] U. D. Hanebeck and V. Klumpp. Localized Cumulative Distributions and a Multivariate Generalization of the Cramér-von Mises Distance. In *Proceedings of the 2008 IEEE International Conference on Multisensor Fusion and Integration for Intelligent Systems (MFI 2008)*, pages 33–39, Seoul, Republic of Korea, Aug. 2008.

[45] C. S. Herz. Bessel Functions of Matrix Argument. *Annals of Mathematics*, 61(3):474–523, May 1955.

[46] M. Huber, P. Krauthausen, and U. D. Hanebeck. Superficial Gaussian Mixture Reduction. In *Proceedings of the IEEE ISIF Workshop on Sensor Data Fusion: Trends, Solutions, Applications (SDF 2011)*, Berlin, Germany, Oct. 2011.

[47] M. F. Huber and U. D. Hanebeck. Gaussian Filter based on Deterministic Sampling for High Quality Nonlinear Estimation. In *Proceedings of the 17th IFAC World Congress (IFAC 2008)*, volume 17, Seoul, Republic of Korea, July 2008.

[48] K. Ito and K. Xiong. Gaussian Filters for Nonlinear Filtering Problems. *IEEE Transactions on Automatic Control*, 45(5):910–927, May 2000.

[49] S. R. Jammalamadaka and A. Sengupta. *Topics in Circular Statistics.* World Scientific Pub Co Inc, 2001.

[50] A. H. Jazwinski. *Stochastic Processes and Filtering Theory.* Dover Publications, 1970.

[51] R. A. Johnson and T. Wherly. Measures and Models for Angular Correlation and Angular-Linear Correlation. *Journal of the Royal Statistical Society. Series B (Methodological)*, 39(2), 1977.

[52] S. Julier. The Spherical Simplex Unscented Transformation. In *Proceedings of the 2003 American Control Conference, 2003.*, volume 3, pages 2430–2434. IEEE, 2003.

[53] S. Julier and J. Uhlmann. Unscented Filtering and Nonlinear Estimation. *Proceedings of the IEEE*, 92(3):401 – 422, 2004.

[54] S. Julier, J. Uhlmann, and H. Durrant-Whyte. A New Approach for Filtering Nonlinear Systems. In *Proceedings of 1995 American Control Conference - ACC'95*, volume 3, pages 1628–1632. American Autom Control Council, 1995.

[55] S. Julier, J. Uhlmann, and H. Durrant-Whyte. A New Method for the Nonlinear Transformation of Means and Covariances in Filters and Estimators. *IEEE Transactions on Automatic Control*, 45(3):477–482, Mar. 2000.

[56] S. J. Julier and J. J. LaViola, Jr. On Kalman Filtering With Nonlinear Equality Constraints. *IEEE Transactions on Signal Processing*, 55(6):2774–2784, June 2007.

[57] A. M. Kagan, Y. V. Linnik, and C. R. Rao. *Characterization Problems in Mathematical Statistics.* Wiley, 1973.

[58] G. Kallianpur. *Stochastic filtering theory.* Springer, 1980.

[59] R. E. Kalman. A New Approach to Linear Filtering and Prediction Problems. *Journal of Basic Engineering*, 82(1):35, Mar. 1960.

[60] R. E. Kalman and R. S. Bucy. New Results in Linear Filtering and Prediction Theory. *Journal of Basic Engineering*, 83(1):95, Mar. 1961.

[61] L. Kavan, S. Collins, J. Žára, and C. O'Sullivan. Skinning with Dual Quaternions. In *Proceedings of the 2007 symposium on Interactive 3D graphics and games - I3D '07*, page 39, New York, New York, USA, Apr. 2007. ACM Press.

[62] L. Kavan, S. Collins, J. Žára, and C. O'Sullivan. Geometric Skinning with Approximate Dual Quaternion Blending. *ACM Transactions on Graphics*, 27(4):1–23, Oct. 2008.

[63] I. Kazakov. An Approximate Method for the Statistical Investigation of Nonlinear Systems. *Transactions of Zhukovsky All-Union Higher Aviation Institute*, 1954.

[64] I. Kazakov. Approximate Probabilistic Analysis of the Accuracy of Operation of Essentially Nonlinear Systems. *Avtomatika i Telemekhanika*, 17(5), 1956.

[65] K. Kim and G. Shevlyakov. Why Gaussianity? *IEEE Signal Processing Magazine*, 25(2):102 –113, Mar. 2008.

[66] P. Koev and A. Edelman. The Efficient Evaluation of the Hypergeometric Function of a Matrix Argument. *Math. Comp*, 75:833–846, 2005.

[67] A. Kolmogorov. Sur l'interpolation et Extrapolation des Suites Stationnaires. *Comptes Rendus de l'Académie des Sciences*, 208, 1939.

[68] A. Kolmogorov. Interpolation and Extrapolation. *Dokl. Akad. Nauk SSSR, Ser. Math.*, 5, 1941.

[69] T. Koyama, H. Nakayama, K. Nishiyama, and N. Takayama. Holonomic Gradient Descent for the Fisher–Bingham Distribution on the d-dimensional Sphere. *Computational Statistics*, 29(3-4):661–683, Oct. 2013.

[70] P. Krauthausen and U. D. Hanebeck. Regularized Non-Parametric Multivariate Density and Conditional Density Estimation. In *Proceedings of the 2010 IEEE International Conference on Multisensor Fusion and Integration for Intelligent Systems (MFI 2010)*, Salt Lake City, Utah, USA, Sept. 2010.

[71] M. G. Krein. On a Generalization of Some Investigations of G. Szegö, W. M. Smirnov, and A. N. Kolmogorov. *Dokl. Akad. Nauk SSSR*, 46, 1945.

[72] M. G. Krein. On a Problem of Extrapolation of A. N. Kolmogorov. *Dokl. Akad. Nauk SSSR*, 46, 1945.

[73] J. B. Kuipers. *Quaternions and Rotation Sequences: A Primer with Applications to Orbits, Aerospace and Virtual Reality.* Princeton University Press, 2002.

[74] S. Kullback and R. A. Leibler. On Information and Sufficiency. *The Annals of Mathematical Statistics*, 22(1):79–86, Mar. 1951.

[75] A. Kume, S. P. Preston, and A. T. A. Wood. Saddlepoint Approximations for the Normalizing Constant of Fisher-Bingham Distributions on Products of Spheres and Stiefel Manifolds. *Biometrika*, 100(4):971–984, Aug. 2013.

[76] A. Kume and A. T. Wood. On the Derivatives of The Normalising Constant of the Bingham Distribution. *Statistics & Probability Letters*, 77(8):832–837, Apr. 2007.

[77] A. Kume and A. T. A. Wood. Saddlepoint Approximations for the Bingham and Fisher-Bingham Normalising Constants. *Biometrika*, 92(2):465–476, June 2005.

[78] G. Kurz. *Directional Estimation for Robotic Beating Heart Surgery.* PhD thesis, Karlsruhe Institute of Technology (KIT), 2015.

[79] M. Lang, O. Dunkley, and S. Hirche. Gaussian Process Kernels for Rotations and 6D Rigid Body Motions. In *2014 IEEE International Conference on Robotics and Automation (ICRA)*, pages 5165–5170. IEEE, May 2014.

[80] M. Lang and W. Feiten. MPG - Fast Forward Reasoning on 6 DOF Pose Uncertainty. In *7th German Conference on Robotics (ROBOTIK 2012)*, pages 1–6, Munich, 2012.

[81] J. LaViola. A Comparison of Unscented and Extended Kalman Filtering for Estimating Quaternion Motion. In *Proceedings of the 2003 American Control Conference*, volume 3, pages 2435–2440, 2003.

[82] T. Lefebvre, H. Bruyninckx, and J. De Schutter. Comment on "A New Method for the Nonlinear Transformation of Means and Covariances in Filters and Estimators". *IEEE Transactions on Automatic Control*, 47(8):1406–1409, Aug. 2002.

[83] T. Lefebvre, H. Bruyninckx, and J. De Schutter. *Nonlinear Kalman Filtering for Force-Controlled Robot Tasks.* Springer, 2005.

[84] M. Liggins, I. Kadar, M. Alford, V. Vannicola, and S. Thomopoulos. Distributed Fusion Architectures and Algorithms for Target Tracking. *Proceedings of the IEEE*, 85(1):95–107, Jan. 1997.

[85] J. S. Liu and R. Chen. Sequential Monte Carlo Methods for Dynamic Systems. *Journal of the American Statistical Association*, 93(443):1032–1044, 1998.

[86] J. Lo and A. Willsky. Estimation for Rotational Processes with One Degree of Freedom–Part I: Introduction and Continuous-Time Processes. *IEEE Transactions on Automatic Control*, 20(1):10–21, Feb. 1975.

[87] K. Mardia. Directional Statistics in Geosciences. *Communications in Statistics - Theory and Methods*, 10(15):1523–1543, June 2007.

[88] K. V. Mardia. Characterizations of Directional Distributions. In *A Modern Course on Statistical Distributions in Scientific Work*, pages 365–385. Springer, 1975.

[89] K. V. Mardia and P. E. Jupp. *Directional Statistics.* Wiley, 1st edition, 1999.

[90] J. Marins, E. Bachmann, R. McGhee, and M. Zyda. An Extended Kalman Filter for Quaternion-based Orientation Estimation Using MARG Sensors. In *Proceedings 2001 IEEE/RSJ International Conference on Intelligent Robots and Systems*, volume 4, pages 2003–2011. IEEE, 2001.

[91] R. Martinez-Cantin and J. Castellanos. Unscented SLAM for Large-Scale Outdoor Environments. In *2005 IEEE/RSJ International Conference on Intelligent Robots and Systems*, pages 3427–3432, 2005.

[92] G. Matsuda, S. Kaji, and H. Ochiai. Anti-commutative Dual Complex Numbers and 2D Rigid Transformation. In *Mathematical Progress in Expressive Image Synthesis I*, pages 131–138. Springer, 2014.

[93] F. Mirzaei and S. Roumeliotis. A Kalman Filter-Based Algorithm for IMU-Camera Calibration: Observability Analysis and Performance Evaluation. *IEEE Transactions on Robotics*, 24(5):1143–1156, Oct. 2008.

[94] R. Muirhead. *Aspects of Multivariate Statistical Theory.* Wiley series in probability and mathematical statistics: Probability and mathematical statistics. Wiley, 1982.

[95] H. Nakayama, K. Nishiyama, M. Noro, K. Ohara, T. Sei, N. Takayama, and A. Takemura. Holonomic Gradient Descent and its Application to the Fisher–Bingham Integral. *Advances in Applied Mathematics*, 47(3):639–658, Sept. 2011.

[96] F. W. J. Olver, D. W. Lozier, R. F. Boisvert, and C. W. Clark. *NIST Handbook of Mathematical Functions.* Cambridge University Press, 2010.

[97] R. L. Plackett. Some Theorems in Least Squares. *Biometrika*, 37(1-2), 1950.

[98] S. T. Rachev and L. Rüschendorf. Probability Metrics and Recursive Algorithms. *Advances in Applied Probability*, 27(3):770–799, 1995.

[99] M. Reinhardt, B. Noack, and U. D. Hanebeck. The Hypothesizing Distributed Kalman Filter. In *Proceedings of the 2012 IEEE International Conference on Multisensor Fusion and Integration for Intelligent Systems (MFI 2012)*, Hamburg, Germany, Sept. 2012.

[100] M. Reinhardt, B. Noack, and U. D. Hanebeck. Advances in Hypothesizing Distributed Kalman Filtering. In *Proceedings of the 16th International Conference on Information Fusion (Fusion 2013)*, Istanbul, Turkey, July 2013.

[101] A. Rényi. On Measures of Entropy and Information. In *Proceedings of the Fourth Berkeley Symposium on Mathematical Statistics and Probability, Volume 1: Contributions to the Theory of Statistics.* The Regents of the University of California, 1961.

[102] L.-P. Rivest. Some Statistical Methods for Bivariate Circular Data. *Journal of the Royal Statistical Society. Series B (Methodological)*, 44(1):81–90, 1982.

[103] X. Rong Li and V. Jilkov. Survey of Maneuvering Target Tracking. Part I: Dynamic Models. *IEEE Transactions on Aerospace and Electronic Systems*, 39(4):1333–1364, Oct. 2003.

[104] A. M. Sabatini. Quaternion-based Extended Kalman Filter for Determining Orientation by Inertial and Magnetic Sensing. *IEEE transactions on bio-medical engineering*, 53(7):1346–56, July 2006.

[105] S. Särkkä. *Bayesian Filtering and Smoothing.* Cambridge University Press, 2013.

[106] W. Schmidt. Statistische Methoden beim Gefügestudium Krystalliner Schiefer. In *Sitzungsberichte Akademie der Wissenschaften in Wien*, 1917.

[107] R. Schubert, N. Mattern, and G. Wanielik. An Evaluation of Nonlinear Filtering Algorithms for Integrating GNSS and Inertial Measurements. In *2008 IEEE/ION Position, Location and Navigation Symposium*, pages 25–29. IEEE, 2008.

[108] A. N. Shiryaev. *Probability.* Springer, 2nd edition, 1995.

[109] K. Shoemake. Animating Rotation with Quaternion Curves. *ACM SIGGRAPH Computer Graphics*, 19(3):245–254, July 1985.

[110] D. Simon. *Optimal State Estimation.* Wiley-Interscience, 2006.

[111] G. L. Smith, S. F. Schmidt, and L. A. McGee. Application of Statistical Filter Theory to the Optimal Estimation of Position and Velocity on Board a Circumlunar Vehicle. Technical report, National Aeronautics and Space Administration (NASA), 1962.

[112] H. Sorenson and D. Alspach. Recursive Bayesian Estimation Using Gaussian Sums. *Automatica*, 7(4):465–479, July 1971.

[113] J. Steinbring and U. D. Hanebeck. S2KF: The Smart Sampling Kalman Filter. In *Proceedings of the 16th International Conference on Information Fusion (Fusion 2013)*, Istanbul, Turkey, July 2013.

[114] R. F. Stengel. *Optimal Control and Estimation.* Dover Publications, 1994.

[115] O. Straka, J. Dunik, and M. Simandl. Randomized Unscented Kalman filter in Target Tracking. In *Proceedings of the 15th International Conference on Information Fusion (Fusion 2012)*, pages 503–510, 2012.

[116] Y. S. Suh. Orientation Estimation Using a Quaternion-Based Indirect Kalman Filter With Adaptive Estimation of External Acceleration. *IEEE Transactions on Instrumentation and Measurement*, 59(12):3296–3305, Dec. 2010.

[117] F. Thomas. Approaching Dual Quaternions From Matrix Algebra. *IEEE Transactions on Robotics*, 30(5):1–12, 2014.

[118] R. Van Der Merwe. *Sigma-Point Kalman Filters for Probabilistic Inference in Dynamic State-Space Models.* Phd thesis, Oregon Health and Science University, Jan. 2004.

[119] R. Van Der Merwe, E. A. Wan, and S. Julier. Sigma-Point Kalman Filters for Nonlinear Estimation and Sensor-Fusion: Applications to Integrated Navigation. In *Proceedings of the AIAA Guidance, Navigation & Control Conference*, Providence, RI, USA, 2004.

[120] R. von Mises. Über die "Ganzzahligkeit" der Atomgewichte und verwandte Fragen. *Physikalische Zeitschrift*, XIX, 1918.

[121] M. W. Walker, L. Shao, and R. A. Volz. Estimating 3-D Location Parameters Using Dual Number Quaternions. *CVGIP: Image Understanding*, 54(3):358–367, Nov. 1991.

[122] X. Wang and D. B. Dunson. Parallelizing MCMC via Weierstrass Sampler. Dec. 2013.

[123] L. N. Wasserstein. Markov Processes Over Denumerable Products of Spaces, Describing Large Systems of Automata. *Problems of Information Transmission*, 5(3):47–52, 1969.

[124] N. Wiener. *Extrapolation, Interpolation, and Smoothing of Stationary Time Series.* Wiley, New York, NY, 1949.

[125] X. Yun, M. Lizarraga, E. Bachmann, and R. McGhee. An Improved Quaternion-based Kalman Filter for Real-Time Tracking of Rigid Body Orientation. In *Proceedings 2003 IEEE/RSJ International Conference on Intelligent Robots and Systems (IROS 2003)*, volume 2, pages 1074–1079. IEEE, 2003.

[126] A. J. Zeevi and R. Meir. Density Estimation Through Convex Combinations of Densities: Approximation and Estimation Bounds. *Neural Networks*, 10(1):99–109, Jan. 1997.

[127] V. M. Zolotarev. Probability Metrics. *Teoriya Veroyatnostei i ee Primeneniya*, 28(1):278–302, 1983.

Supervised Student Theses

[128] S. Dingler, A.-M. Hellmund, and A. Weigl. *Orientierungsschätzung mobiler Geräte mit der Bingham-Verteilung.* Student research project, Intelligent Sensor-Actuator-Systems Laboratory, Karlsruhe Institute of Technology (KIT), 2014.

[129] H. Klein, N. Wagensommer, and J. Walk. IEEE Signal Processing Cup 2015. Student research project, Intelligent Sensor-Actuator-Systems Laboratory, Karlsruhe Institute of Technology (KIT), 2015.

[130] F. Mabrouk. *Filter and Portfolio Optimization for Unobservable Markov-Modulated Stockdrift and Discrete Observations of the Stock-prices.* Diploma thesis, Karlsruhe Institute of Technology (KIT), 2013.

[131] M. Sons. *A Probabilistic Approach to Sensor Management for Dynamic Systems.* Master thesis, Intelligent Sensor-Actuator-Systems Laboratory, Karlsruhe Institute of Technology (KIT), 2013.

[132] M. Sons, D. Maucher, and R. Sandkühler. *Construction and Control of a Flying Ball.* Lab project, Intelligent Sensor-Actuator-Systems Laboratory, Karlsruhe Institute of Technology (KIT), 2012.

[133] H. Stenger. *Mode-S-Radar-Monitor.* Diploma thesis, Intelligent Sensor-Actuator-Systems Laboratory, Karlsruhe Institute of Technology (KIT), 2014.

Own Publications

[134] I. Gilitschenski and U. D. Hanebeck. A Robust Computational Test for Overlap of Two Arbitrary-dimensional Ellipsoids in Fault-Detection of Kalman Filters. In *Proceedings of the 15th International Conference on Information Fusion (Fusion 2012)*, Singapore, July 2012.

[135] I. Gilitschenski and U. D. Hanebeck. Efficient Deterministic Dirac Mixture Approximation. In *Proceedings of the 2013 American Control Conference (ACC 2013)*, Washington D. C., USA, June 2013.

[136] I. Gilitschenski and U. D. Hanebeck. A Direct Method for Checking Overlap of Two Hyperellipsoids. In *Proceedings of the IEEE ISIF Workshop on Sensor Data Fusion: Trends, Solutions, Applications (SDF 2014)*, Bonn, Germany, Oct. 2014.

[137] I. Gilitschenski, G. Kurz, and U. D. Hanebeck. Bearings-Only Sensor Scheduling Using Circular Statistics. In *Proceedings of the 16th International Conference on Information Fusion (Fusion 2013)*, Istanbul, Turkey, July 2013.

[138] I. Gilitschenski, G. Kurz, S. J. Julier, and U. D. Hanebeck. A New Probability Distribution for Simultaneous Representation of Uncertain Position and Orientation. In *Proceedings of the 17th International Conference on Information Fusion (Fusion 2014)*, Salamanca, Spain, July 2014.

[139] I. Gilitschenski, G. Kurz, S. J. Julier, and U. D. Hanebeck. Efficient Bingham Filtering based on Saddlepoint Approximations. In *Proceedings of the 2014 IEEE International Conference on Multisensor Fusion and Information Integration (MFI 2014)*, Beijing, China, Sept. 2014.

[140] I. Gilitschenski, G. Kurz, S. J. Julier, and U. D. Hanebeck. Unscented Orientation Estimation Based on the Bingham Distribution. *IEEE Transactions on Automatic Control (accepted)*, 2015.

[141] I. Gilitschenski, J. Steinbring, U. D. Hanebeck, and M. Simandl. Deterministic Dirac Mixture Approximation of Gaussian Mixtures. In *Proceedings of the 17th International Conference on Information Fusion (Fusion 2014)*, Salamanca, Spain, July 2014.

[142] G. Kurz, I. Gilitschenski, M. Dolgov, and U. D. Hanebeck. Bivariate Angular Estimation Under Consideration of Dependencies Using Directional Statistics. In *Proceedings of the 53rd IEEE Conference on Decision and Control (CDC 2014)*, Los Angeles, California, USA, Dec. 2014.

[143] G. Kurz, I. Gilitschenski, and U. D. Hanebeck. Recursive Nonlinear Filtering for Angular Data Based on Circular Distributions. In *Proceedings of the 2013 American Control Conference (ACC 2013)*, Washington D. C., USA, June 2013.

[144] G. Kurz, I. Gilitschenski, and U. D. Hanebeck. Deterministic Approximation of Circular Densities with Symmetric Dirac Mixtures Based on Two Circular Moments. In *Proceedings of the 17th International Conference on Information Fusion (Fusion 2014)*, Salamanca, Spain, July 2014.

[145] G. Kurz, I. Gilitschenski, and U. D. Hanebeck. Efficient Evaluation of the Probability Density Function of a Wrapped Normal Distribution. In *Proceedings of the IEEE ISIF Workshop on Sensor Data Fusion: Trends, Solutions, Applications (SDF 2014)*, Bonn, Germany, Oct. 2014.

[146] G. Kurz, I. Gilitschenski, and U. D. Hanebeck. Nonlinear Measurement Update for Estimation of Angular Systems Based on Circular Distributions. In *Proceedings of the 2014 American Control Conference (ACC 2014)*, Portland, Oregon, USA, June 2014.

[147] G. Kurz, I. Gilitschenski, and U. D. Hanebeck. The Partially Wrapped Normal Distribution for SE(2) Estimation. In *Proceedings of the 2014 IEEE International Conference on Multisensor Fusion and Information Integration (MFI 2014)*, Beijing, China, Sept. 2014.

[148] G. Kurz, I. Gilitschenski, S. Julier, and U. D. Hanebeck. Recursive Bingham Filter for Directional Estimation Involving 180 Degree Symmetry. *Journal of Advances in Information Fusion*, 9(2):90 – 105, Dec. 2014.

[149] G. Kurz, I. Gilitschenski, S. J. Julier, and U. D. Hanebeck. Recursive Estimation of Orientation Based on the Bingham Distribution. In *Proceedings of the 16th International Conference on Information Fusion (Fusion 2013)*, Istanbul, Turkey, July 2013.

Karlsruhe Series on Intelligent Sensor-Actuator-Systems

Edited by Prof. Dr.-Ing. Uwe D. Hanebeck // ISSN 1867-3813

Band 1 **Oliver Schrempf**
Stochastische Behandlung von Unsicherheiten in kaskadierten dynamischen Systemen. 2008
ISBN 978-3-86644-287-0

Band 2 **Florian Weißel**
Stochastische modell-prädiktive Regelung nichtlinearer Systeme. 2009
ISBN 978-3-86644-348-8

Band 3 **Patrick Rößler**
Telepräsente Bewegung und haptische Interaktion in ausgedehnten entfernten Umgebungen. 2009
ISBN 978-3-86644-346-4

Band 4 **Kathrin Roberts**
Modellbasierte Herzbewegungsschätzung für robotergestützte Interventionen. 2009
ISBN 978-3-86644-353-2

Band 5 **Felix Sawo**
Nonlinear state and parameter estimation of spatially distributed systems. 2009
ISBN 978-3-86644-370-9

Band 6 **Gregor F. Schwarzenberg**
Untersuchung der Abbildungseigenschaften eines 3D-Ultraschall-Computertomographen zur Berechnung der 3D-Abbildungsfunktion und Herleitung einer optimierten Sensorgeometrie. 2009
ISBN 978-3-86644-393-8

Karlsruhe Series on Intelligent Sensor-Actuator-Systems

Edited by Prof. Dr.-Ing. Uwe D. Hanebeck // ISSN 1867-3813

Band 7	**Marco Huber** Probabilistic Framework for Sensor Management. 2009 ISBN 978-3-86644-405-8
Band 8	**Frederik Beutler** Probabilistische modellbasierte Signalverarbeitung zur instantanen Lageschätzung. 2010 ISBN 978-3-86644-442-3
Band 9	**Marc Peter Deisenroth** Efficient Reinforcement Learning using Gaussian Processes. 2010 ISBN 978-3-86644-569-7
Band 10	**Evgeniya Ballmann** Physics-Based Probabilistic Motion Compensation of Elastically Deformable Objects. 2012 ISBN 978-3-86644-862-9
Band 11	**Peter Krauthausen** Learning Dynamic Systems for Intention Recognition in Human-Robot-Cooperation. 2013 ISBN 978-3-86644-952-7
Band 12	**Antonia Pérez Arias** Haptic Guidance for Extended Range Telepresence. 2013 ISBN 978-3-7315-0035-3
Band 13	**Marcus Baum** Simultaneous Tracking and Shape Estimation of Extended Objects. 2013 ISBN 978-3-7315-0078-0

Die Bände sind unter www.ksp.kit.edu als PDF frei verfügbar oder als Druckausgabe bestellbar.

Karlsruhe Series on Intelligent Sensor-Actuator-Systems

Edited by Prof. Dr.-Ing. Uwe D. Hanebeck // ISSN 1867-3813

Band 14 **Benjamin Noack**
State Estimation for Distributed Systems with Stochastic and Set-membership Uncertainties. 2014
ISBN 978-3-7315-0124-4

Band 15 **Jörg Fischer**
Optimal Sequence-Based Control of Networked Linear Systems. 2014
ISBN 978-3-7315-0305-7

Band 16 **Marc Reinhardt**
Linear Estimation in Interconnected Sensor Systems with Information Constraints. 2015
ISBN 978-3-7315-0342-2

Band 17 **Gerhard Kurz**
Directional Estimation for Robotic Beating Heart Surgery. 2015
ISBN 978-3-7315-0382-8

Band 18 **Igor Gilitschenski**
Deterministic Sampling for Nonlinear Dynamic State Estimation. 2016
ISBN 978-3-7315-0473-3

Die Bände sind unter www.ksp.kit.edu als PDF frei verfügbar oder als Druckausgabe bestellbar.